Statistical Analysis with R

Beginner's Guide

Take control of your data and produce superior statistical analyses with R

John M. Quick

BIRMINGHAM - MUMBAI

Statistical Analysis with R
Beginner's Guide

First published: October 2010

Production Reference: 1191010

Published by Packt Publishing Ltd.
32 Lincoln Road
Olton
Birmingham, B27 6PA, UK.

ISBN 978-1-849512-08-4

www.packtpub.com

Cover Image by John M. Quick (john@johnmquick.com)

Credits

Author
John M. Quick

Reviewers
Ajay Ohri
Joshua Wiley

Acquisition Editor
Douglas Paterson

Development Editor
Meeta Rajani

Technical Editor
Vanjeet D'souza

Indexer
Tejal Daruwale

Editorial Team Leader
Akshara Aware

Project Team Leader
Priya Mukherji

Project Coordinator
Jovita Pinto

Proofreaders
Aaron Nash
Chris Smith

Graphics
Nilesh Mohite

Production Coordinator
Aparna Bhagat

Cover Work
Aparna Bhagat

About the Author

John M. Quick is an Educational Technology Ph.D. student at Arizona State University who is interested in the design, research, and use of educational innovations. Currently, his work focuses on mixed-reality systems, interactive media, and innovation adoption. In addition, he has recently published multiple gaming applications for the iPhone and iPad. John's blog, **High-Technically Correct**, which covers various topics in technology, is available online at `http://www.johnmquick.com`.

I give thanks to the R Project and its user community for offering the world superior open-source statistical software. I also thank Dr. Roy Levy for introducing me to, and encouraging me to share my knowledge of, R. Lastly, I would like to thank my parents for their lifelong support and Zarraz for the companionship and insights that she offered to me throughout the authoring of this book.

About the Reviewers

Ajay Ohri has been working in the field of analytics since 2004 , when it was a still nascent emerging Industry in India. He has worked with the top two Indian outsourcers listed on NYSE, and with Citigroup on cross-sell analytics where he helped sell an extra 50000 credit cards by cross-sell analytics .He was one of the very first independent data mining consultants in India working on analytics products and domestic Indian market analytics. He regularly writes on analytics topics on his website `www.decisionstats.com` and is currently working on open source analytical tools like R and analytical software like SAS.

Joshua Wiley has implemented R in several laboratories on multiple campuses of the University of California system to run statistical analyses and produce high-quality graphics. He also uses it for data processing in descriptive and inferential statistics. He is currently working towards his Ph.D. at UCLA, where he researches Health Psychology. In addition to his own work with R, Mr. Wiley has led tutorials for other psychology researchers on using R, and is an active member of the R-help mailing list.

Table of Contents

Preface 1

Chapter 1: Uncovering the Strategist's Data Analysis Tool 7

What is R? 8

What are the benefits of using R? 8

Why should I use R? 9

Why should I read this book? 9

What topics are covered in this book? 9

Chapter 2—Preparing R for Battle 10

Chapter 3—Exploring the Mysterious Data Analysis Tool 11

Chapter 4—Collecting and Organizing Information 11

Chapter 5—Assessing the Situation 12

Chapter 6—Planning the Attack 12

Chapter 7—Organizing the Battle Plans 13

Chapter 8—Briefing the Emperor 14

Chapter 9—Briefing the Generals 15

Chapter 10—Becoming a Master Strategist 17

Summary 17

Chapter 2: Preparing R for Battle 19

Time for action – downloading and installing R 20

Example: R 2.11.1 Mac OS X 10.5+ installation wizard demonstration 24

Time for action – issuing your first R command 29

Time for action – setting your R working directory 30

Summary 32

Chapter 3: Exploring the Mysterious Data Analysis Tool 33

Deciphering Zhuge Liang's magic square 34

Time for action – solving the first 4x4 magic square 35

Lines 37

Comments 37

Calculations 38
Output 38
Visualizing the R console 39
Summary **41**

Chapter 4: Collecting and Organizing Information **43**
Time for action – importing external data **43**
read.csv(file) 44
comma-separated values (csv) files 44
Time for action – creating and calling variables **45**
Time for action – accessing data within variables **47**
variable$column notation 49
attach(variable) function 49
variable[row, column] notation 50
Time for action – manipulating variable data **51**
Performing a calculation on an entire dataset 53
Performing a calculation on a row, column, or cell 54
Using variable data in function arguments 54
Saving a variable calculation into a new variable 55
Time for action – managing the R workspace **57**
Listing the contents of the R workspace 58
Saving the contents of the R workspace 59
Loading the contents of the R workspace 59
Quitting R 59
Distinguishing between the R console and workspace 59
Saving the R console 60
Summary **62**

Chapter 5: Assessing the Situation **63**
Time for action – making an initial inference from our data **63**
Examining our data **65**
Time for action – creating a subset from a large dataset **66**
Multi-argument functions 67
Variable-argument functions 67
Equivalency operators 67
subset(data, ...) 67
Time for action – deriving summary statistics **69**
Means 71
Standard deviations 71
Ranges 72
summary(object) 72
Why use summary statistics? 72

Time for action – quantifying categorical variables **73**
as.numeric(data) 75
Overwriting variables 75
Time for action – correlating variables **77**
Interpreting correlations 78
cor(x, y) 79
cor(data) 80
NA values 80
Regression **82**
Time for action – modelling with simple linear regression **82**
lm(formula, data) 84
Linear model output 84
Linear model summary 85
Interpreting a linear regression model 86
Time for action – modelling with multiple linear regression **88**
Interpreting the summary output 90
Explaining model differences 91
Time for action – modelling interactions **92**
Interpreting interaction variables 94
Time for action – comparing and choosing models **96**
Interpreting the model summaries 98
Interpreting the ANOVA results 99
anova(object, ...) 100
Summary **101**
Chapter 6: Planning the Attack **103**
Review of models **103**
Head to head 104
Surround 105
Ambush 106
Fire 107
Predicting outcomes using regression models **108**
Rating 108
Successfully executed 108
Number of Wei soldiers 109
Duration of battle 110
A word about assumptions 110
Time for action – calculating outcomes from regression models **110**
Time for action – creating custom functions **111**
function() 113
Extended lines 114

Time for action – creating resource-focused custom functions **115**
Logistical considerations **117**
Gold 117
Provisions 117
Equipment 118
Soldiers 118
Resource and cost summary 118
Resource map 118
Time for action – incorporating resource constraints into predictions **119**
Gold cost function explanation 120
Assessing viability **121**
Time for action – assessing the viability of potential strategies **122**
Remember your assumptions 122
Summary **124**
Chapter 7: Organizing the Battle Plans **125**
Retracing and refining a complete analysis **125**
Time for action – first steps **126**
Time for action – data setup **126**
read.table(...) 128
Time for action – data exploration **129**
Time for action – model development **132**
glm(...) 138
AIC(object, ...) 138
Time for action – model deployment **139**
coef(object) 143
Time for action – last steps **145**
The common steps to all R analyses **145**
Step 1: Set your working directory 145
Comment your work 146
Step 2: Import your data (or load an existing workspace) 146
Step 3: Explore your data 147
Step 4: Conduct your analysis 148
Step 5: Save your workspace and console files 148
Summary **150**
Chapter 8: Briefing the Emperor **151**
Charts, graphs, and plots in R **151**
Time for action – creating a bar chart **152**
barplot(...) 153
Vectors 154
Graphic window 154

Time for action – customizing graphics **156**
Graphic customization arguments 159
main, xlab, and ylab 159
xlim and ylim 160
Col 161
legend(...) 162
Time for action – creating a scatterplot **164**
Single scatterplot 167
Multiple scatterplots 167
Time for action – creating a line chart **168**
type 170
Number-colon-number notation 170
Time for action – creating a box plot **172**
boxplot(...) 174
Time for action – creating a histogram **175**
hist(...) 176
Time for action – creating a pie chart **177**
pie(...) 179
Time for action – exporting graphics **181**
Summary **184**
Chapter 9: Briefing the Generals **185**
More charts, graphs, and plots in R **186**
Time for action – customizing a bar chart **186**
names 194
width and space 194
horiz 195
beside 196
density and angle 197
legend(...) with density, angle, and cex 198
Time for action – customizing a scatterplot **199**
pch and cex 206
points(...) 207
legend(...) 209
abline(...) 209
Time for action – customizing a line chart **212**
lwd 216
lines(...) 217
legend(...) 219
Time for action – customizing a box plot **220**
range 223
axis(...) 223

Time for action – customizing a histogram **225**
breaks 228
freq 228
Time for action – customizing a pie chart **230**
Custom labels 231
legend(...) 233
Time for action – building a graphic **234**
Time for action – building a graphic with multiple visuals **242**
par(mfcol) 249
Graphics 249
Horizontal and vertical lines 250
Nested functions 250
Summary **252**
Chapter 10: Becoming a Master Strategist **253**
R's built-in resources **253**
Time for action – using R's help function **254**
help(...) 256
Time for action – expanding R with packages **257**
Choose a CRAN mirror 260
Install a package 260
Load the package 260
Use the package 261
R's online resources **262**
Websites 263
The R Project for Statistical Computing 263
Quick-R 263
R Programming wikibook 263
R Graph Gallery 263
Crantastic! 264
Blogs 264
R bloggers 264
R Tutorial Series 264
Online communities 264
R-help mailing list 264
Other mailing lists 265
Search engines 265
R Seek 265
Google 265
Summary **266**

Appendix: Pop Quiz Answer Key **267**
Chapter 2 267
Chapter 3 267
Chapter 4 267
Chapter 5 268
Chapter 6 269
Chapter 7 270
Chapter 8 270
Chapter 9 271
Chapter 10 273
Index **275**

Preface

You have unexpectedly been thrust into the role of lead strategist for the kingdom. After you install your predecessor's mysterious data analysis tool, you will begin to explore its fundamental elements. Next, you will use R to import and organize your data. Then, you will use functions and statistical analyses to arrive at potential courses of action. Subsequently, you will design your own functions to assess the practical impacts of your predictions. Lastly, you will focus on communicating your results through the use of charts, plots, graphs, and custom built visualizations. The fate of the kingdom is in your hands. Your rapid development as a master R strategist is the key to future success.

What this book covers

Chapter 1, Uncovering the Strategist's Data Analysis Tool, serves as an introduction to the R Project. We will explore the benefits of using R and the topics covered in this book.

Chapter 2, Preparing R for Battle, includes a step-by-step guide to downloading and installing R. We will also launch R and execute our first commands.

Chapter 3, Exploring the Mysterious Data Analysis Tool, is an introduction to the R interface and programming language. In this chapter, we will use R to solve a complex puzzle.

Chapter 4, Collecting and Organizing Information, covers how to import data into R and manipulate it using variables. We will also learn how manage the R workspace.

Chapter 5, Assessing the Situation, focuses on evaluating our data and using it to generate predictive models. We will also consider the statistical and practical significance of our analyses.

Chapter 6, Planning the Attack, involves using our data models to predict potential outcomes and assess their logistical viability. Along the way, we will learn to build our own custom functions.

Chapter 7, Organizing the Battle Plans, revisits the task of planning and organizing a complete data analysis, such that it can be effectively communicated to others. Throughout this process, we will apply the common steps to all R analyses.

Chapter 8, Briefing the Emperor, is a first look at R's graphical capabilities. We will make customizable charts, graphs, and plots that can be exported for use outside of R.

Chapter 9, Briefing the Generals, examines the in-depth customization options available to several types of charts, graphs, and plots. We will also build our own custom graphics from scratch.

Chapter 10, Becoming a Master Strategist, describes the resources that are available to you beyond the contents of this book for further expanding your knowledge of R.

What you need for this book

This code used in this book should be applicable to any version of R on any platform, although it was generated and tested using R 2.11.1 for Mac OS X.

Who this book is for

You want to take control of your data and learn how to conduct effective analyses with R. Whether you are a data analyst, business or information technology professional, student, educator, researcher, or anyone else who wants to learn about R, this book is for you.

No prior experience with R is necessary. Knowledge of other programming languages, software packages, or statistics may be helpful, but is not required. With a willingness to learn and an interest in conducting superior data analyses, you will quickly become an experienced and knowledgeable R user.

Conventions

In this book, you will find several headings appearing frequently.

To give clear instructions of how to complete a procedure or task, we use:

Time for action – heading

1. Action 1
2. Action 2
3. Action 3

Instructions often need some extra explanation so that they make sense, so they are followed with:

What just happened?

This heading explains the working of tasks or instructions that you have just completed.

You will also find some other learning aids in the book, including:

Pop quiz—heading

These are short multiple choice questions intended to help you test your own understanding.

Have a go hero—heading

These set practical challenges and give you ideas for experimenting with what you have learned.

You will also find a number of styles of text that distinguish between different kinds of information. Here are some examples of these styles, and an explanation of their meaning.

Code words in text are shown as follows: "We also expanded upon the `legend(...)` function to gain more control over its appearance."

A block of code is set as follows:

```
> barplot(height = barAllMethodsDurationBars,
main = barAllMethodsDurationLabelMain,
xlab = barAllMethodsDurationLabelX,
ylab = barAllMethodsDurationLabelY,
xlim = barAllMethodsDurationLimX,
ylim = barAllMethodsDurationLimY,
col = barAllMethodsDurationRainbowColors)
```

When we wish to draw your attention to a particular part of a code block, the relevant lines or items are set in bold:

```
> barplot(height = barAllMethodsDurationBars,
main = barAllMethodsDurationLabelMain,
xlab = barAllMethodsDurationLabelY,
ylab = barAllMethodsDurationLabelX,
xlim = barAllMethodsDurationLimY,
ylim = barAllMethodsDurationLimX,
col = barAllMethodsDurationRainbowColors)
```

New terms and **important words** are shown in bold. Words that you see on the screen, in menus or dialog boxes for example, appear in the text like this: "The **R Help** window will open to display documentation on the provided function".

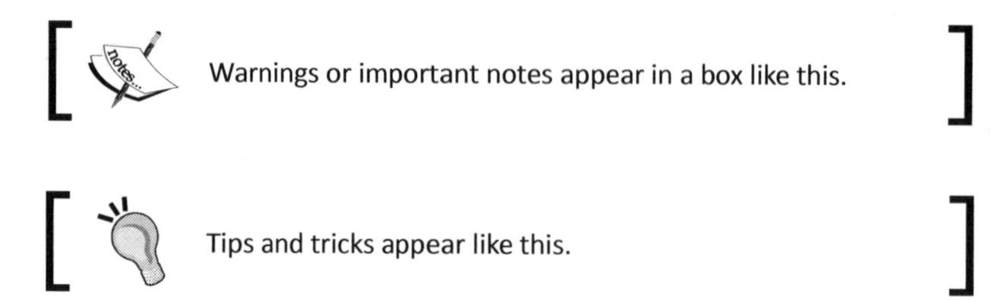

[Warnings or important notes appear in a box like this.]

[Tips and tricks appear like this.]

Reader feedback

Feedback from our readers is always welcome. Let us know what you think about this book—what you liked or may have disliked. Reader feedback is important for us to develop titles that you really get the most out of.

To send us general feedback, simply send an e-mail to `feedback@packtpub.com`, and mention the book title via the subject of your message.

If there is a book that you need and would like to see us publish, please send us a note in the **SUGGEST A TITLE** form on `www.packtpub.com` or e-mail `suggest@packtpub.com`.

If there is a topic that you have expertise in and you are interested in either writing or contributing to a book, see our author guide on `www.packtpub.com/authors`.

Customer support

Now that you are the proud owner of a Packt book, we have a number of things to help you to get the most from your purchase.

Downloading the example code for this book

You can download the example code files for all Packt books you have purchased from your account at `http://www.PacktPub.com`. If you purchased this book elsewhere, you can visit `http://www.PacktPub.com/support` and register to have the files e-mailed directly to you.

Errata

Although we have taken every care to ensure the accuracy of our content, mistakes do happen. If you find a mistake in one of our books—maybe a mistake in the text or the code—we would be grateful if you would report this to us. By doing so, you can save other readers from frustration and help us improve subsequent versions of this book. If you find any errata, please report them by visiting `http://www.packtpub.com/support`, selecting your book, clicking on the **errata submission form** link, and entering the details of your errata. Once your errata are verified, your submission will be accepted and the errata will be uploaded on our website, or added to any list of existing errata, under the Errata section of that title. Any existing errata can be viewed by selecting your title from `http://www.packtpub.com/support`.

Piracy

Piracy of copyright material on the Internet is an ongoing problem across all media. At Packt, we take the protection of our copyright and licenses very seriously. If you come across any illegal copies of our works, in any form, on the Internet, please provide us with the location address or website name immediately so that we can pursue a remedy.

Please contact us at `copyright@packtpub.com` with a link to the suspected pirated material.

We appreciate your help in protecting our authors, and our ability to bring you valuable content.

Questions

You can contact us at `questions@packtpub.com` if you are having a problem with any aspect of the book, and we will do our best to address it.

1

Uncovering the Strategist's Data Analysis Tool

Near the end of the second century A.D., China's Han dynasty crumbled and left numerous warlords fighting for the throne. By the start of the third century, three kingdoms—Shu, Wei, and Wu—emerged as contenders for China's rule. These factions would vie for power for the better part of 80 years during what is known as the Three Kingdoms period of Chinese history.

The most famous military strategist of the era, Zhuge Liang, joined the Shu army in 207 A.D. He is well known for baffling opposing forces with ingenious techniques and cunning tactics. As a result, Zhuge Liang remains a Chinese cultural symbol of intellect and wisdom to this day. In 228 A.D., Zhuge Liang would launch the first of five campaigns against the rival kingdom of Wei. During his fifth, and final, campaign at the Wuzhang Plains, Zhuge Liang fell terminally ill. Following his death in August of 234 A.D., the Shu army was forced to withdraw from its conflict with the kingdom of Wei.

— Taken from Three Kingdoms. Beijing, China: Foreign Language Press; Luo Guanzhong. Translator Moss Roberts.

Prior to his passing, the legendary strategist chose you to succeed him as commander of the Shu forces. Zhuge Liang also left you with secret documents that reveal the knowledge of a powerful data analysis tool.

With your forces currently recuperating in Hanzhong, China, it is your duty to plan the next move. Armed with the late strategist's tool and your talents for data analysis, the fate of the Shu kingdom is in your hands.

By the end of this chapter, you will be able to:

- Describe the R Project for Statistical Computing
- Detail how you will benefit from using R
- Explain why R is an essential tool for your work
- Decide why this book is right for you
- List the major topics covered in this book

What is R?

As the newly appointed strategist for the Shu army, your decisions will impact the lives of many. Great decisions tend not to occur by random chance. Rather, they are a product of knowledge, planning, and sound rationale. A major factor in generating fruitful outcomes is considering the available information and using it to assess your potential courses of action. Fortunately, an essential software tool exists that will help you rise to the occasion and make the most of any situation.

The **R Project for Statistical Computing** (or just **R** for short) is a powerful data analysis tool. It is both a programming language and a computational and graphical environment.

R is free, open source software made available under the GNU General Public License. It runs on Mac, Windows, and Unix operating systems.

The official R website is available at the following site:

`http://www.r-project.org`

What are the benefits of using R?

There are several ways in which R will benefit you, be it as an information technology professional, business analyst, leader of the Shu army, or otherwise. These benefits are discussed in the following points:

- **Free**: R is available to you at no cost. The saying, "give a person a data analysis tool and he or she will learn to analyze data" has never been more true.
- **Cross-platform**: R runs on Mac, Windows, and numerous Unix systems. Whether you are visiting the Emperor in Chengdu or laying siege to the enemy capital at Luoyang, you can be confident that your software will run, regardless of the local operating system.
- **Open source**: R is open source. It allows you to exercise your genius in ways that a closed software does not.

- **Programmable**: R includes a powerful yet straightforward programming language that is designed to compliment the formation of complex strategies.
- **Extendable**: R can be expanded through thousands of available packages. If you are looking for a function to calculate the odds of a successful fire attack, the chances are someone has already made it. If not, you can create it and offer it to the world.
- **Graphical**: R contains robust graphical capabilities. Whether you are looking to create an unassuming plot of provision use over time or an elaborate array of battle maps, R is at your service.
- **Community-supported**: R has a vast user community that is continually updating and contributing to its capabilities. Even the great Zhuge Liang had to rely on his allies from time to time.

Why should I use R?

You should use R because you are interested in taking control of and making the most out of your data. R provides you with opportunities to design and execute complex, customized analyses that other software packages do not. At the same time, R remains accessible and relevant to a large audience of potential users.

With the fate of a kingdom resting upon your shoulders, you can ill afford a miscalculation or misinterpretation. R will assist you in making the best possible decisions and allow you to rise to greatness as a premier strategist.

Why should I read this book?

You should read this book because you are interested in learning how to improve your work through the use of R. You do not need to be an expert at using a programming language, other software packages, or statistics. No prior experience with R is necessary. With a willingness to learn and an interest in conducting superior data analyses, you will quickly become an experienced and knowledgeable user of R.

What topics are covered in this book?

This book covers an extensive range of topics in R. It will comfortably and rapidly familiarize you with the basics, before you proceed into in-depth analyses and custom graphics. A brief description of each chapter's content is provided.

Chapter 2—Preparing R for Battle

In this chapter, we will step through the R installation process. Afterwards, you will launch R and execute your first commands in the R console.

By the end of the chapter, you will be able to:

- Download R
- Install R
- Run R on your computer
- Issue an R command
- Set your R working directory

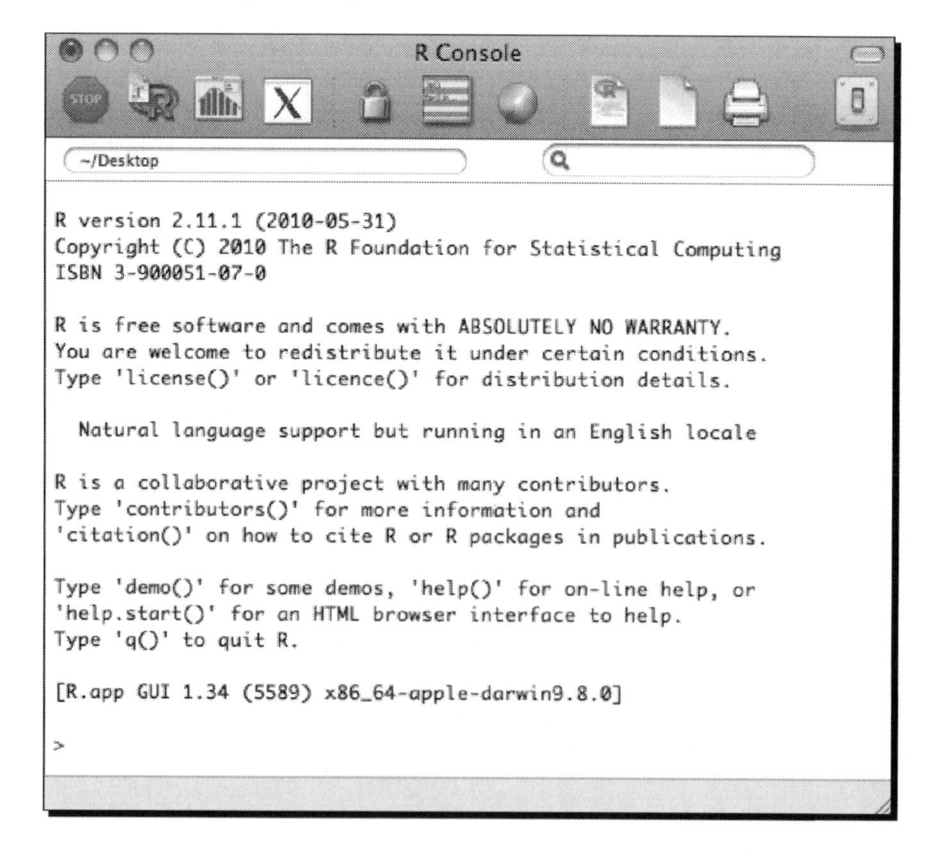

Chapter 3—Exploring the Mysterious Data Analysis Tool

In this chapter, we will explore the anatomy of the R console in greater depth by solving a challenging puzzle that was presented to us by the late Zhuge Liang.

By the end of the chapter, you will be able to:

- Use proper syntax within the R console
- Comment your R code
- Make calculations using formulas
- Distinguish between different types of input and output in the R console

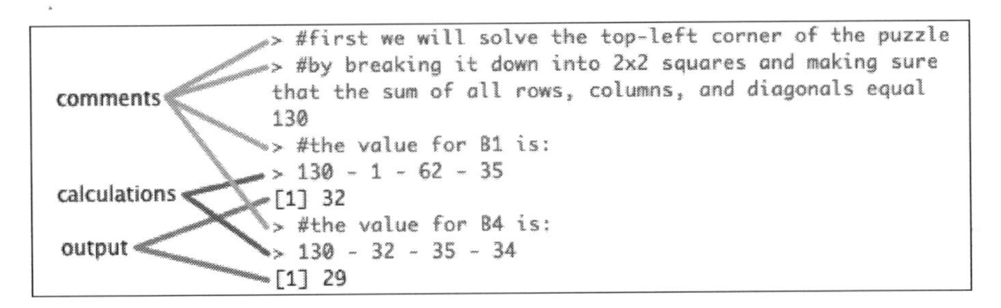
comments
calculations
output
> #first we will solve the top-left corner of the puzzle
> #by breaking it down into 2x2 squares and making sure that the sum of all rows, columns, and diagonals equal 130
> #the value for B1 is:
> 130 - 1 - 62 - 35
[1] 32
> #the value for B4 is:
> 130 - 32 - 35 - 34
[1] 29

Chapter 4—Collecting and Organizing Information

In this chapter, we will focus on getting our data into R and then manipulating it via variables. We will also learn how to manage the R workspace.

By the end of the chapter, you will be able to:

- Import external data into R
- Use variables to organize and manipulate your data
- Manage the R workspace

```
> ls()
[1] "baxiSoldiersAfterRelocation"
[2] "guanghanSoldiersAfterRelocation"
[3] "hanzhongResources"
[4] "hanzhongResourcesAfterFlood"
[5] "maxSoldiersByCity"
[6] "meanSoldiersByCity"
[7] "minSoldiersByCity"
[8] "soldiersByCity"
[9] "totalSoldiers"
```

Chapter 5—Assessing the Situation

In this chapter, we will extensively examine and evaluate our data. This will entail the use of diverse functions to create predictive data models. Throughout this process, we will also consider the practical and statistical meaning behind our analyses.

By the end of the chapter, you will be able to:

- Use multi-argument and variable-argument functions to make calculations
- Create predictive data models using regression analysis
- Consider the statistical and practical significance of your analyses

```
> subsetHeadToHead
       Method Rating SuccessfullyExecuted  Result ShuSoldiersEngaged WeiSoldiersEngaged DurationInDays
1  headToHead      5                    Y  Defeat               5000              15000            116
2  headToHead     15                    Y  Defeat               5000              10000             96
3  headToHead     25                    Y  Defeat               5000              10000             76
4  headToHead     25                    Y  Defeat               5000              10000             61
5  headToHead     35                    Y  Defeat               7500              15000             52
6  headToHead     20                    Y  Defeat              50000             100000             94
7  headToHead     30                    Y  Defeat             100000             200000             87
8  headToHead     85                    Y Victory              10000               5000             36
9  headToHead     95                    Y Victory             100000              50000             44
10 headToHead     85                    Y Victory              30000              15000             49
11 headToHead     10                    Y  Defeat               2500               5000            112
12 headToHead     15                    Y  Defeat               2000               2500             90
13 headToHead     15                    Y  Defeat                250                500            111
14 headToHead     20                    Y  Defeat               1000               2000             93
15 headToHead     10                    Y  Defeat               7000               7500            120
16 headToHead     10                    Y  Defeat               5000               7500            100
17 headToHead     90                    Y Victory              15000              10000             35
18 headToHead     80                    Y Victory              15000              10000             45
19 headToHead     85                    Y Victory              25000              10000             40
20 headToHead     85                    Y Victory              25000              20000             45
21 headToHead     35                    Y  Defeat              30000              35000             95
22 headToHead     45                    Y  Defeat              25000              35000            105
23 headToHead     50                    Y  Defeat              40000              45000            100
24 headToHead     35                    Y  Defeat              30000              45000             91
25 headToHead     25                    Y  Defeat              65000              75000            120
26 headToHead     20                    Y  Defeat              50000              75000             99
27 headToHead     30                    Y  Defeat              60000              75000            102
28 headToHead     90                    Y Victory              75000              40000             44
29 headToHead     90                    Y Victory              50000              25000             50
30 headToHead     95                    Y Victory             100000              60000             30
```

Chapter 6—Planning the Attack

In this chapter, we will turn towards using our data models to predict outcomes. We will also assess the viability of these outcomes. Along the way, we will create and employ our own custom functions that expand the capabilities of R.

By the end of the chapter, you will be able to:

- Use regression models to predict outcomes
- Create your own custom functions to address specific needs
- Assess the viability of achieving the outcomes predicted by regression models.

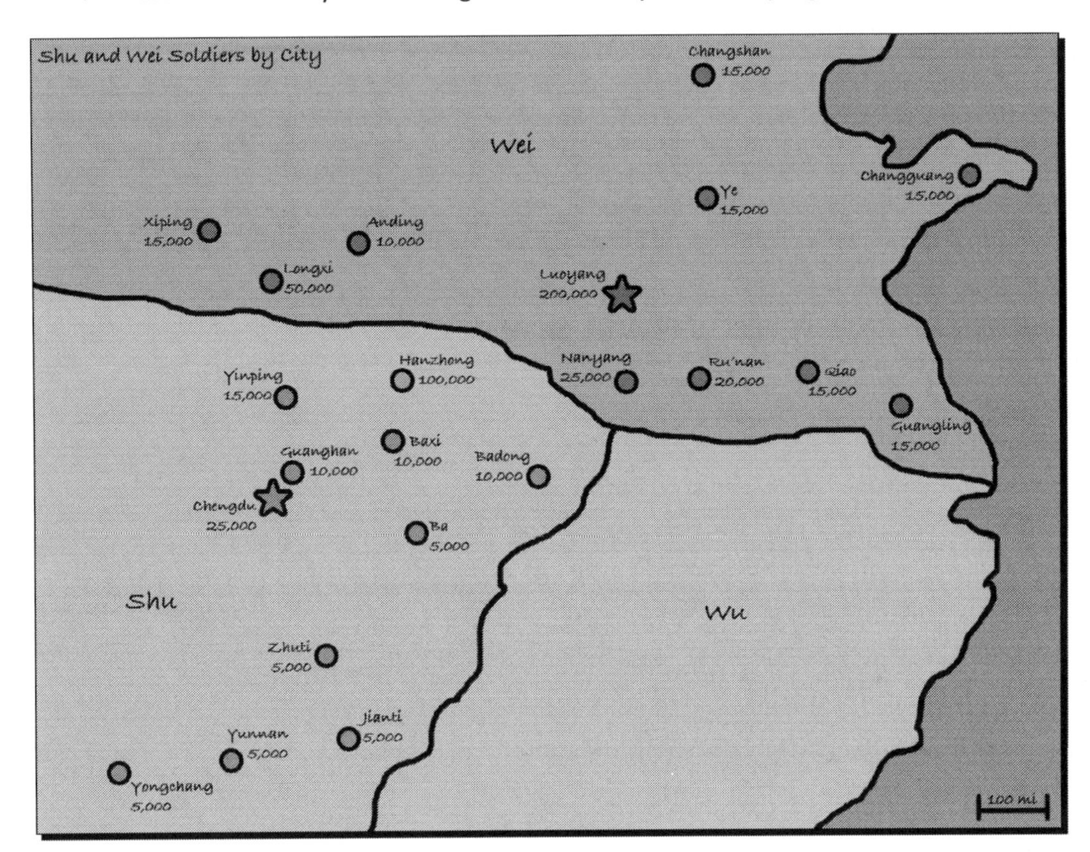

Chapter 7—Organizing the Battle Plans

In this chapter, our task will be to review and organize a complete data analysis. We will emphasize the need to clarify and communicate our data analyses effectively, which can be achieved through a series of common steps.

By the end of the chapter, you will be able to:

- Organize and clarify your raw R data analyses
- Communicate your raw R data analyses in the most effective manner
- Apply the steps common to all well-conducted R analyses

```
> lmFireRating_ExecutionDuration_Summary

Call:
glm(formula = Rating ~ SuccessfullyExecuted + DurationInDays,
    data = subsetFire)

Deviance Residuals:
    Min       1Q   Median       3Q      Max
-26.515   -6.360    3.052    8.377   12.905

Coefficients:
                     Estimate Std. Error t value Pr(>|t|)
(Intercept)           38.1253     4.0899   9.322 6.27e-10 ***
SuccessfullyExecuted  52.9484     3.9815  13.299 2.28e-13 ***
DurationInDays        -1.6177     0.4493  -3.600  0.00126 **
---
Signif. codes:  0 '***' 0.001 '**' 0.01 '*' 0.05 '.' 0.1 ' ' 1

(Dispersion parameter for gaussian family taken to be 98.24895)

    Null deviance: 25396.7  on 29  degrees of freedom
Residual deviance:  2652.7  on 27  degrees of freedom
AIC: 227.6

Number of Fisher Scoring iterations: 2
```

Chapter 8—Briefing the Emperor

In this chapter, we will take our first look at R's graphical capabilities by generating several charts, graphs, and plots. Throughout, we will use common graphical parameters to customize these visuals. We will also save and export our graphics for external use.

By the end of the chapter, you will be able to:

- Create six different charts, graphs, and plots in R
- Customize your R visuals using text, colors, axes, and legends

- Save and export your graphics for use outside of R

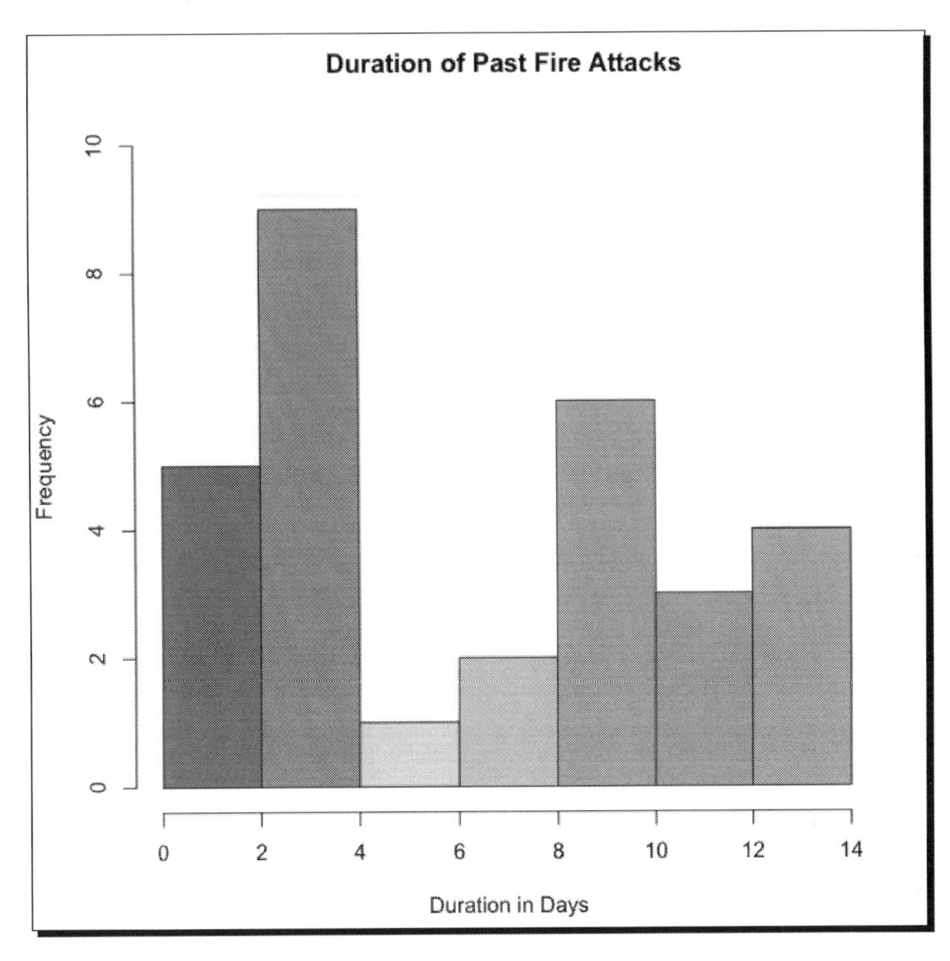

Chapter 9—Briefing the Generals

In this chapter, we will take a deeper look at R's graphical capabilities. We will practice customizing different types of charts, graphs, and plots by modifying their unique parameters. We will also learn how to build our own custom graphics from scratch using R's graphics functions.

By the end of the chapter, you will be able to:

- Customize several charts, graphs, and plots using arguments specific to each
- Use graphics functions to add information to any visual
- Create custom graphics by building them from the ground up

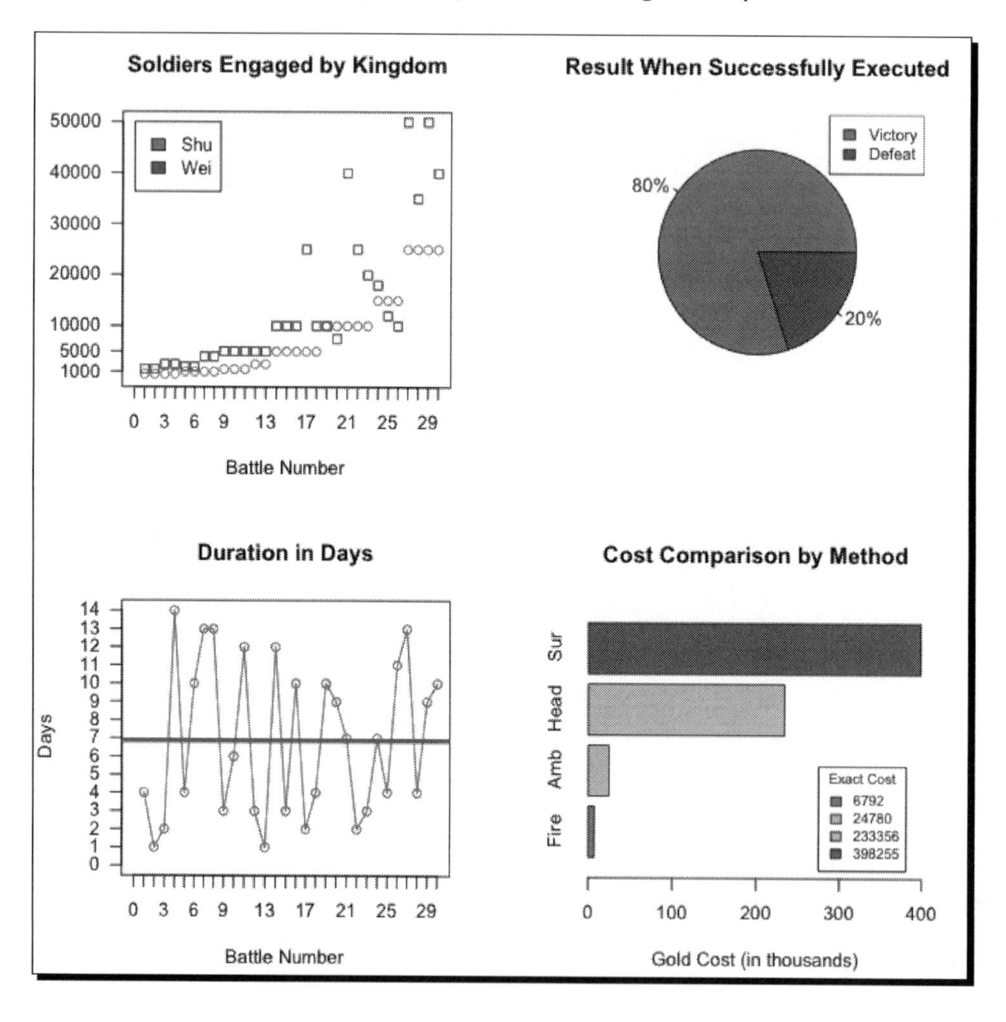

Chapter 10—Becoming a Master Strategist

In the final chapter, we will look to the future. We will focus on the ways in which you can learn beyond the contents of this book to further expand your knowledge of R.

By the end of the chapter, you will be able to:

- Use R's built-in help system
- Install packages that expand R's functionality
- Take advantage of electronic learning resources, such as websites, blogs, and online communities

Summary

In this chapter, we were introduced to R. We learned that its benefits include being free, cross-platform, open source, programmable, extendable, graphical, and community-supported. We also considered why you should use R to conduct your data analyses and how this book can help you quickly become an experienced R user.

You should now be able to:

- Describe the R Project for Statistical Computing
- Detail how you will benefit from using R
- Explain why R is an essential tool for your work
- Decide why this book is right for you
- List the major topics covered in this book

In the next chapter, we will work through the installation process to prepare R for battle.

2
Preparing R for Battle

Before you can begin to formulate a strategy for the Shu forces, you must ensure that your data analysis tool is in working order. Fortunately, R can be prepared for battle in a few straightforward steps.

By the end of this chapter, you will be able to:

- Download R
- Install R
- Run R on your computer
- Issue an R command
- Set your R working directory

Time for action – downloading and installing R

Let us see now how to download and install R:

1. Browse to the official R website at `http://www.r-project.org`; the home page looks like the following:

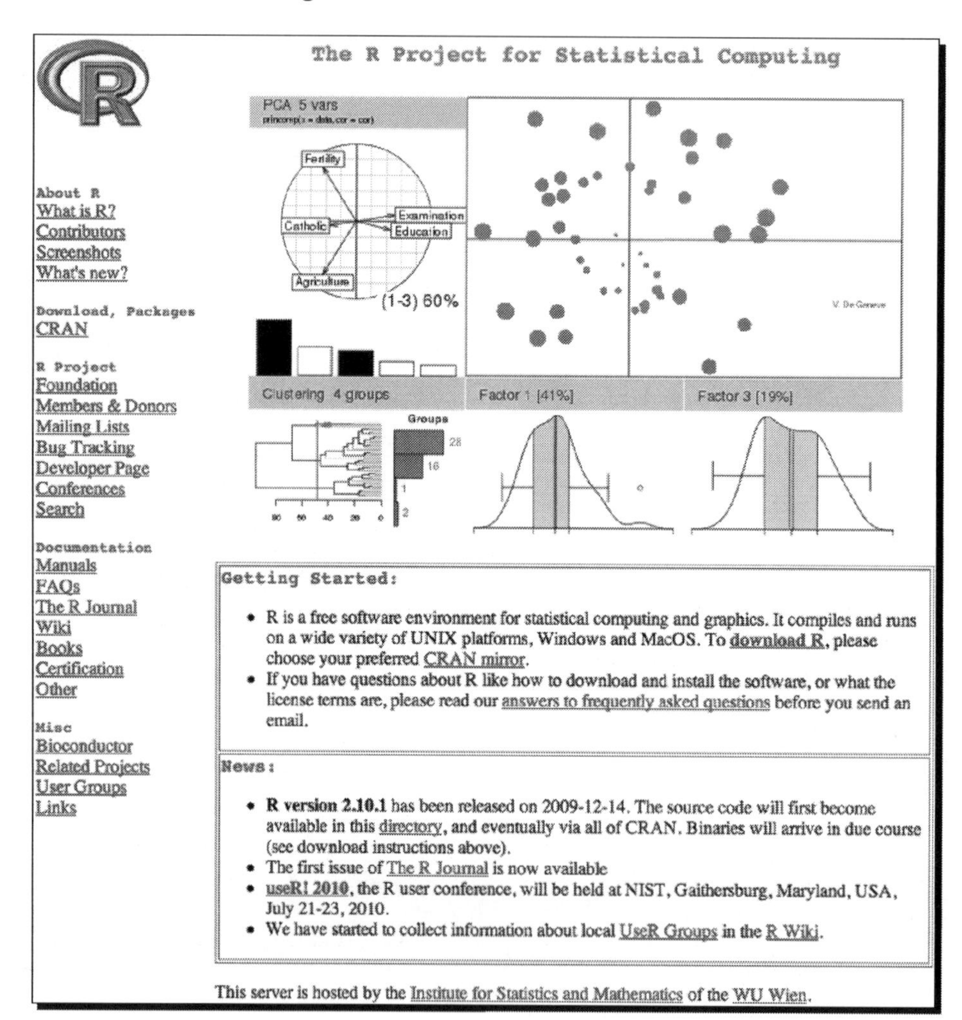

2. Under the **Download, Packages** heading on the left-hand side of the screen, click on the **CRAN** link.

Download, Packages
CRAN

3. A page with several CRAN (Comprehensive R Archive Network) servers located across the world will be displayed. Click on the link for the CRAN server located nearest to you.

About R
What is R?
Contributors
Screenshots
What's new?

Download, Packages
CRAN

R Project
Foundation
Members & Donors
Mailing Lists
Bug Tracking
Developer Page
Conferences
Search

Documentation
Manuals
FAQs
The R Journal
Wiki
Books
Certification
Other

Misc
Bioconductor
Related Projects
User Groups
Links

CRAN Mirrors

The Comprehensive R Archive Network is available at the following URLs, please choose a location close to you. Some statistics on the status of the mirrors can be found here.

Argentina	
http://cran.patan.com.ar/	Patan.com.ar, Buenos Aires
Australia	
http://cran.ms.unimelb.edu.au/	University of Melbourne
Austria	
http://cran.at.r-project.org/	Wirtschaftsuniversitaet Wien
Belgium	
http://www.freestatistics.org/cran/	K.U.Leuven Association
Brazil	
http://cran.br.r-project.org/	Universidade Federal do Parana
http://cran.fiocruz.br/	Oswaldo Cruz Foundation, Rio de Janeiro
http://www.vps.fmvz.usp.br/CRAN/	University of Sao Paulo, Sao Paulo
http://brieger.esalq.usp.br/CRAN/	University of Sao Paulo, Piracicaba
Canada	
http://cran.stat.sfu.ca/	Simon Fraser University, Burnaby
http://probability.ca/cran/	University of Toronto
http://cran.skazkaforyou.com/	iWeb, Montreal
http://cran.parentinginformed.com/	iWeb, Montreal
Chile	
http://dirichlet.mat.puc.cl/	Pontificia Universidad Catolica de Chile, Santiago
China	
http://ftp.ctex.org/mirrors/CRAN/	CTEX.ORG
http://cran.csdb.cn/	Computer Network Information Center, CAS, Beijing
http://mirrors.geoexpat.com/cran/	GeoExpat.Com
Colombia	
http://www.laqee.unal.edu.co/CRAN/	National University of Colombia
Denmark	
http://cran.dk.r-project.org/	dotsrc.org, Aalborg
France	
http://cran.fr.r-project.org/	CICT, Toulouse
http://cran.univ-lyon1.fr/	Dept. of Biometry & Evol. Biology, University of Lyon
http://cran.miroir-francais.fr/	Miroir-Francais, Paris

4. A page with frequently used CRAN links will be displayed. In the **Download and Install R** section, click on the link that corresponds to your operating system (**Linux**, **Mac OS X**, or **Windows**).

The Comprehensive R Archive Network
Frequently used pages
CRAN
Mirrors
What's new?
Task Views
Search
About R
R Homepage
The R Journal
Software
R Sources
R Binaries
Packages
Other
Documentation
Manuals
FAQs
Contributed
Download and Install R
Precompiled binary distributions of the base system and contributed packages, Windows and Mac users most likely want one of these versions of R:
Linux
MacOS X
Windows
Source Code for all Platforms
Windows and Mac users most likely want the precompiled binaries listed in the upper box, not the source code. The sources have to be compiled before you can use them. If you do not know what this means, you probably do not want to do it!
The latest release (2009-12-14): R-2.10.1.tar.gz (read what's new in the latest version).
Sources of R alpha and beta releases (daily snapshots, created only in time periods before a planned release).
Daily snapshots of current patched and development versions are available here. Please read about new features and bug fixes before filing corresponding feature requests or bug reports.
Source code of older versions of R is available here.
Contributed extension packages
Questions About R
If you have questions about R like how to download and install the software, or what the license terms are, please read our answers to frequently asked questions before you send an email.

5. Use the provided link to download the latest version of R for your operating system and version.

> For demonstration purposes, the Mac OS X page is shown here. As of this writing, a user on Mac OS X 10.5 or higher would click on the **R-2.11.1.pkg** link to download the installation package. Similarly, you should download the appropriate installation file for your operating system and version.

R for Mac OS X

This directory contains binaries for a base distribution and packages to run on Mac OS X (release 10.5 and above). Mac OS 8.6 to 9.2 (and Mac OS X 10.1) are no longer supported but you can find the last supported release of R for these systems (which is R 1.7.1) here. Releases for old Mac OS X systems (through Mac OS X 10.4) can be found in the old directory.

CRAN
Mirrors
What's new?
Task Views
Search

About R
R Homepage
The R Journal

Software
R Sources
R Binaries
Packages
Other

Documentation
Manuals
FAQs
Contributed

Note: CRAN does not have Mac OS X systems and cannot check these binaries for viruses. Altough we take precautions when assembling binaries, please use the normal precautions with downloaded executables.

Universal R 2.11.1 released on 2010/05/31

This binary distribution of R and the GUI supports PowerPC (32-bit) and Intel (32-bit and 64-bit) based Macs on Mac OS X 10.5 (Leopard) and 10.6 (Snow Leopard).

Please check the MD5 checksum of the downloaded image to ensure that it has not been tampered with or corrupted during the mirroring process. For example type
`md5 R-2.11.1.pkg`
in the *Terminal* application to print the MD5 checksum for the R-2.11.1.pkg image.

Files:

R-2.11.1.pkg (latest version) MD5-hash: ce4de47e58efb9a69573b86ba0cb5b3d (ca. 38MB)	Three-way universal binary of **R 2.11.1** for Mac OS X 10.5 (Leopard) and higher. Contains R 2.11.1 framework, R.app GUI 1.34 in 32-bit and 64-bit. The above file is an Installer package which can be installed by double-clicking. Depending on your browser, you may need to press the control key and click on this link to download the file.

6. Double-click on the file that you downloaded in step 5. Then follow the prompts to install R on your computer.

> For assistance with your specific operating system, see section **2.5 How can R be installed?** of the official R FAQ at `http://cran.r-project.org/doc/FAQ/R-FAQ.html`. This section provides documentation for installing R on the most frequently used operating systems:
>
> Macintosh: `http://cran.r-project.org/doc/FAQ/R-FAQ.html#How-can-R-be-installed-_0028Macintosh_0029`
>
> Unix-based: `http://cran.r-project.org/doc/FAQ/R-FAQ.html#How-can-R-be-installed-_0028Unix_002dlike_0029`
>
> Windows: `http://cran.r-project.org/doc/FAQ/R-FAQ.html#How-can-R-be-installed-_0028Windows_0029`

Example: R 2.11.1 Mac OS X 10.5+ installation wizard demonstration

For demonstration purposes only, the installation process for `R-2.11.1.pkg` on Mac OS X 10.5 and higher is shown here. The exact installation process will differ between operating systems and versions. Therefore, it is likely that your installation process will differ from the one shown here, although it may also bear some similarities. The process goes as follows:

1. Locate and double-click the `R-2.11.1` package file that you downloaded earlier.

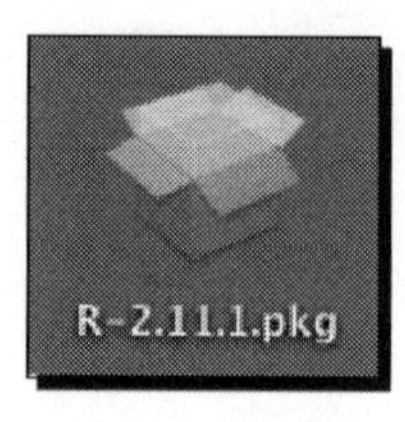

2. The **Install R 2.11.1** wizard will open in a new window. From this **Introduction** page, click on the **Continue** button.

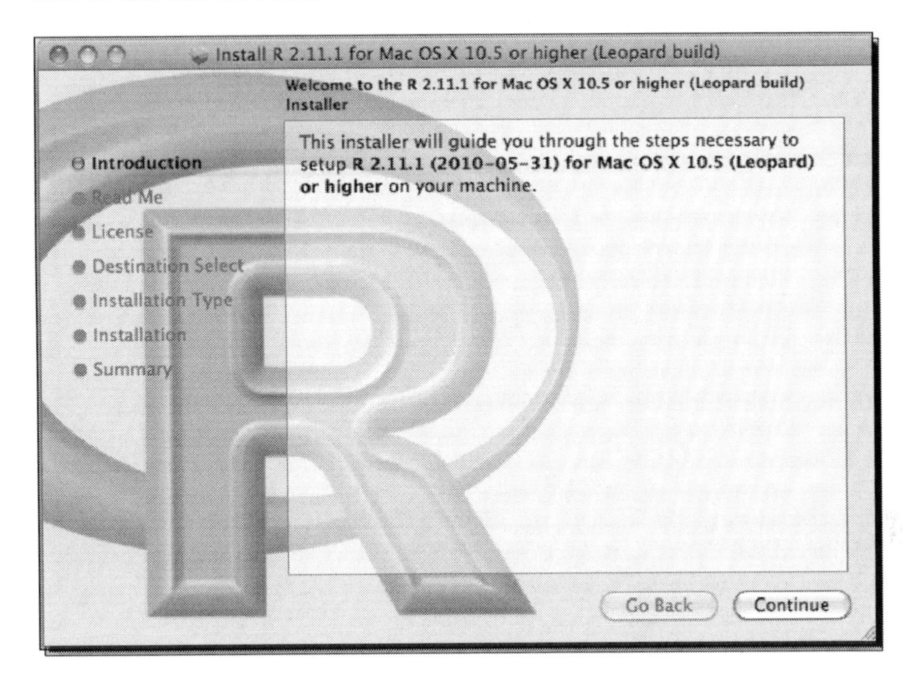

3. The wizard will advance to the **Read Me** page. Click on the **Continue** button.

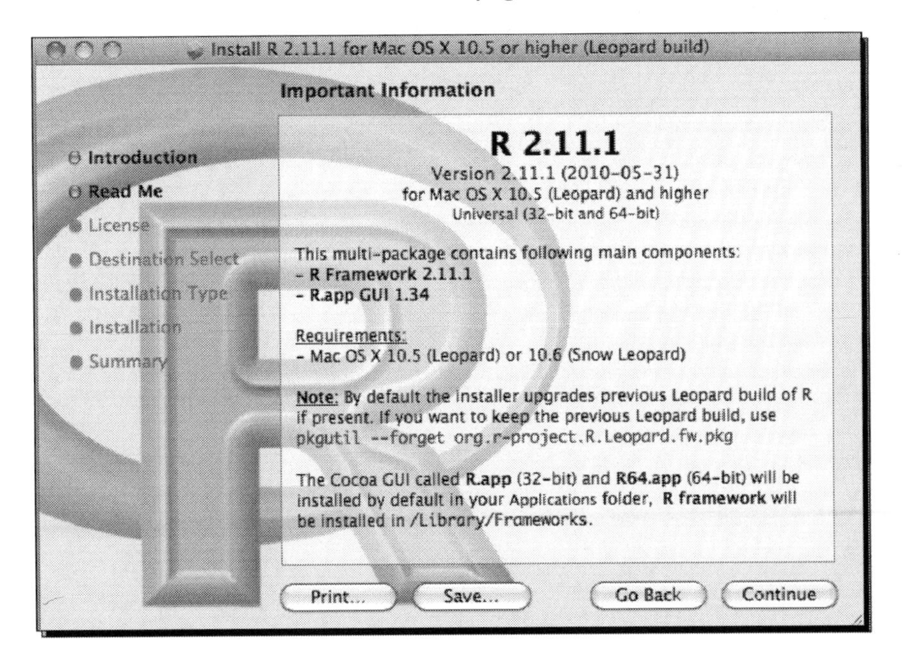

4. The wizard will advance to the **License** page. Click on the **Continue** button.

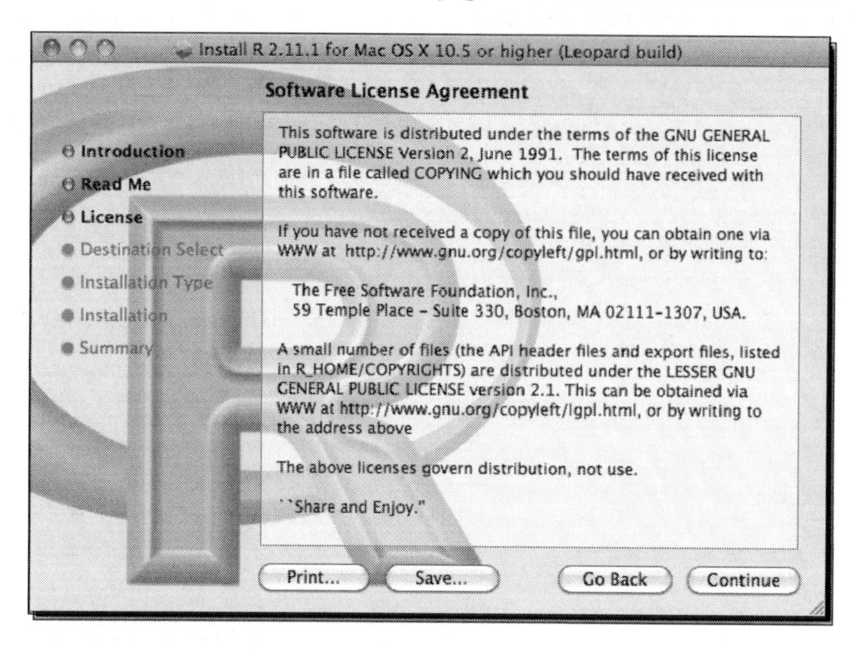

5. A window will pop up asking you to agree to the license terms. Click on the **Agree** button.

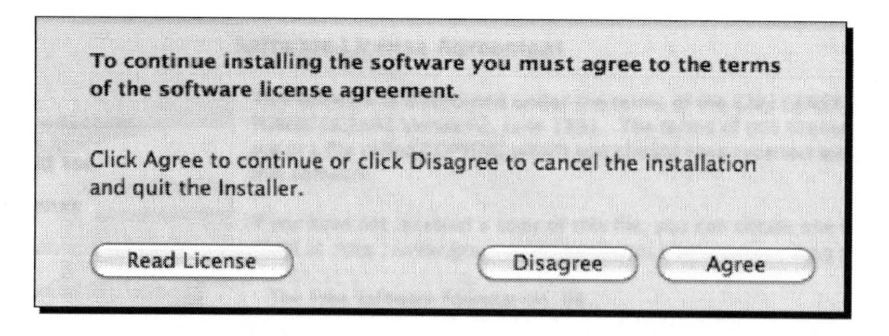

6. The wizard will advance to the **Destination Select** page. Change the installation location, *only* if you have an explicit reason to do so. Otherwise, click on the **Install** button.

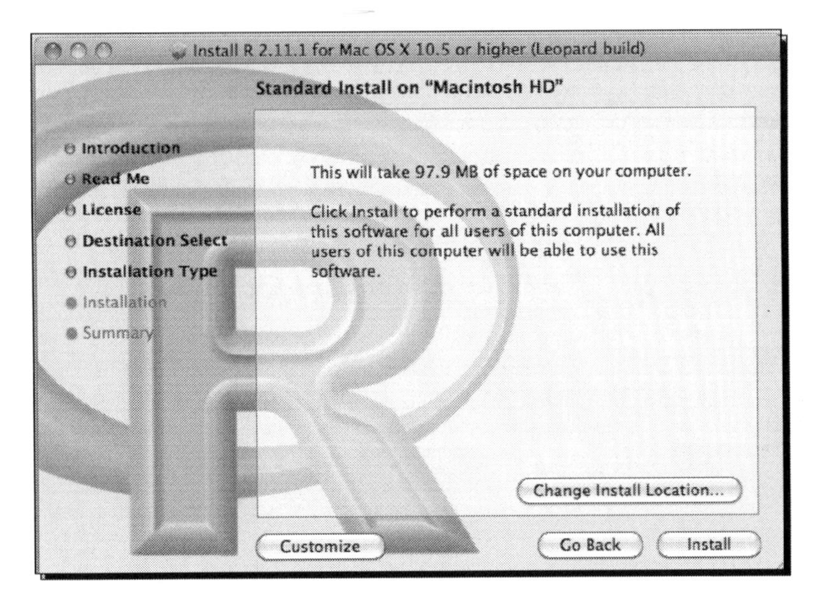

7. The wizard will advance to the **Installation** page. It will automatically install the necessary files on your computer. This process will take approximately five minutes.

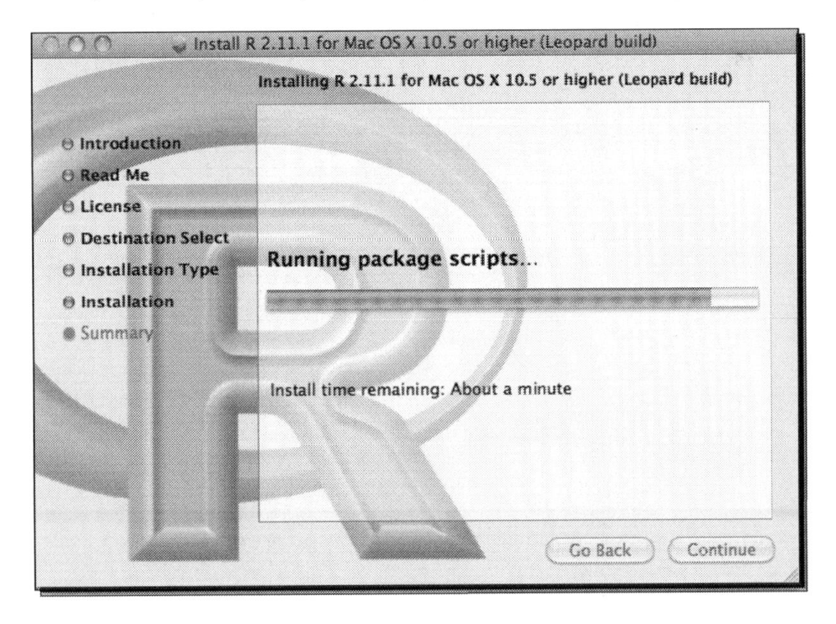

8. Once the installation is complete, the wizard will advance to the **Summary** page. Here, you will receive a message indicating that R was installed successfully. Click on the **Close** button to exit the wizard.

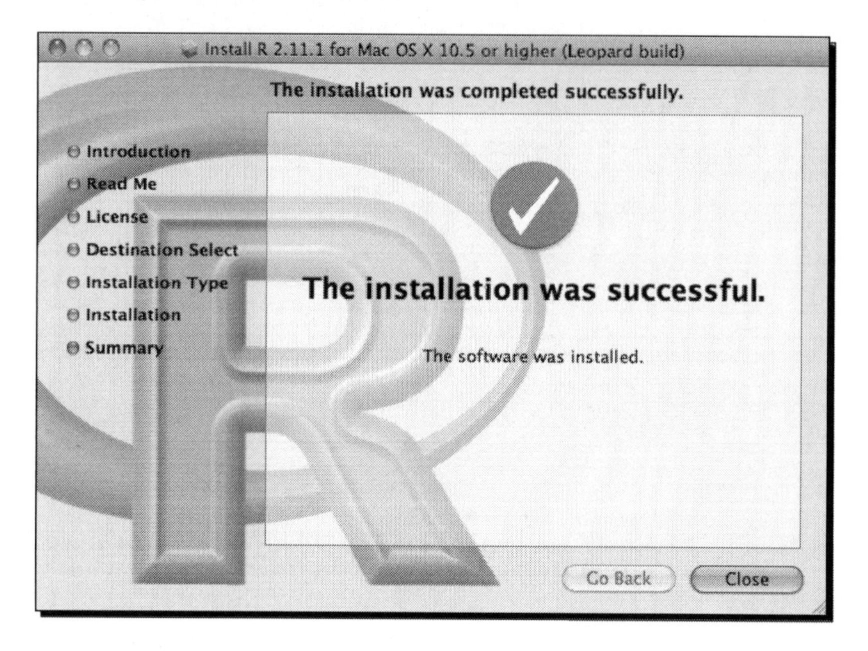

9. You can launch R at any time by browsing to its location on your hard drive and double-clicking on its icon.

10. After completing the installation process, double-click the R icon to launch the R console.

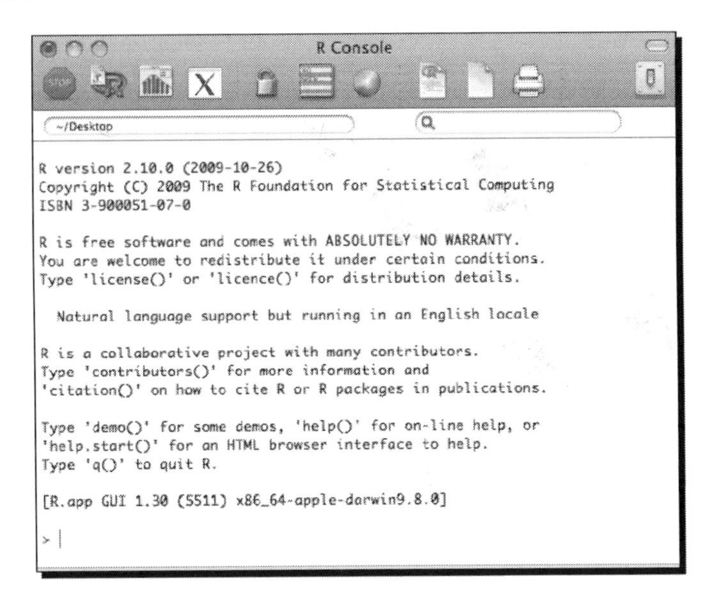

What just happened?

You just finished installing R and launched it for the first time. Next, we will learn how to use R by issuing our first command.

Time for action – issuing your first R command

Time may have escaped you amidst your sudden change in position and hustle in preparing R. Conveniently, R provides us with a simple command to retrieve the current date and time:

1. In the R console, next to the greater than sign (>), type the following comments and then press the *Return* (or *Enter*) key:

```
> #use the date() command to get the current date and time
```

2. Similarly, issue the `date()` command and press the *Return* key:

```
> date()
```

3. After issuing the `date()` command, a message similar to `[1] "Sun Aug 31 08:00:00 234"` will appear in the R console. In this case, the message indicates that it is 8:00 AM on Sunday, August 31 in the year 234.

Congratulations, you have successfully issued your first R command (and reminded yourself of the current date and time in the process).

What just happened?

As you may have noticed, R **commands**, or **functions**, take on a similar form as in other programming languages. In the `date()` command, for example, the word `date` specifies the name of the function, while the parentheses `()` contain the function's arguments. In this case, it was not necessary to use any arguments in the `date()` command. However, other functions, such as `setwd(dir)` receive one or more arguments.

Time for action – setting your R working directory

To demonstrate how to use the `setwd(dir)` and `getwd()` functions, we will set our working directory to the desktop:

1. Determine the file path to your desktop. Note that this path will vary depending on your operating system and the structure of your hard drive. An example, which you should replace with your computer's path, is shown here:

```
> #set the R working directory
> #this should be the first thing you do every time you open R
> #I am going to set the R working directory to my desktop
at "/Users/johnmquick/Desktop"
> #you should replace the sample location with the path to
your desktop
```

2. Use the `setwd(dir)` command to set your working directory to the path identified in step 1:

```
> #use setwd(dir) to set the R working directory
> setwd(dir = "/Users/johnmquick/Desktop")
```

3. Verify that your working directory was set appropriately using the `getwd()` command:

```
> #use getwd() to display the current R working directory
> getwd()
[1] "/Users/johnmquick/Desktop"
```

What just happened?

The **working directory** is the default location on your computer where R assumes all of your work is being conducted at a given time. For example, if you were to import data from an external file, R would automatically look for it in your working directory. Furthermore, all file path arguments in functions are evaluated relative to the working directory. Therefore, it is important to set your working directory each time you use R.

We set our R working directory using the `setwd(dir)` function and then verified its location using the `getwd()` command.

In `setwd(dir)`, the `dir` argument accepts a path to the folder that is to become the working directory. For example, the path `"/Users/johnmquick/Desktop"` tells the `setwd(dir)` function to locate the `Desktop` folder within the `johnmquick` folder of the `Users` folder. It then sets this destination as the working directory.

After submitting the `setwd(dir)` command, R will drop down to the next line without providing any output:

```
> setwd(dir = "/Users/johnmquick/Desktop")
>
```

In one sense, this is good, because you would have received an error if the command failed. However, R can also be confusing at times, because it does not always provide you with feedback indicating the result of your commands.

Note that in our `setwd(dir)` function, the `dir =` portion can be optionally omitted. In R, so long as a function's arguments occur in the default order, they do not have to be explicitly stated in the code. However, if only certain arguments are used, or if they are used in a different order, they must be stated explicitly. We will encountered several examples of both cases throughout this book.

Thankfully, we can use the `getwd()` command to verify the current working directory:

```
> getwd()
[1] "/Users/johnmquick/Desktop"
```

By using `getwd()` after `setwd(dir)`, you can verify that your working directory has been defined appropriately. Remember that setting your working directory is the first thing you should do every time you launch R.

Pop quiz

1. Which of the following is not true of the R working directory?
   a. It is set using the `setwd(dir)` command.
   b. It is displayed using the `getwd()` command.
   c. It is the default location where R assumes your work is being conducted.
   d. It only needs to be set once.

2. Which of the following is true of the R console?
    a. It returns output for no functions entered by the user.
    b. It returns output for some functions entered by the user.
    c. It returns output for all functions entered by the user.
    d. It returns output for all functions, but not comments, entered by the user.
3. In `setwd(dir)`, `dir` is which of the following?
    a. A variable.
    b. A function.
    c. An argument.
    d. An element.

Have a go hero

Set your R working directory to a location of your choice using the `setwd(dir)` function. Then verify the location of your working directory using the `getwd()` command. It may be useful to designate a specific folder for all of your R work or for each individual project that you engage in. For example, you may want to create a specific folder on your computer for all of the activities that we will complete in this book. You could then set that location as your R working directory. Remember that your working directory should be set each time you open R.

Summary

In this chapter, we downloaded, installed, and ran R for the first time. Then, you issued your first R command (of very many to come) and learned how to set and verify the R working directory.

We will begin to explore the mysterious data tool in the following chapter by using it to solve a challenging puzzle. Meanwhile, we will learn about the anatomy of the R console in greater detail.

3
Exploring the Mysterious Data Analysis Tool

With R prepared for use, you are primed to begin your initial status assessment of the Shu army. However, you have realized that the documents that you received make no mention of your own or your enemies' resources. Without this critical data, you will not be able to conduct your analyses.

You decide to pay a visit to Zhuge Liang's assistant to see if the great strategist had mistakenly misplaced the much needed information. Upon your arrival, the assistant silently hands you a written letter. It reads:

> *My true successor will be a person of sharp intellect and patient wisdom. Yet, it is not sufficient to merely choose a replacement. Rather, this person's character must be tested under the harshest of circumstances. I have hidden my records of the Shu and Wei armies. I predicted that you would come for them shortly after my death. My assistant has been instructed to share this letter with you. Further, if you are able to solve the puzzle that I have presented here within one hour, then you will receive what you seek. However, if you cannot complete this task within the given time, the documents will be destroyed and my assistant will promptly travel to the capital. There, my assistant will give the emperor my recommendation that the Shu forces surrender to the Wei kingdom.*
>
> *Zhuge Liang*

1			61		28		57
62	35		2			27	
	34			59	38		7
		33					60
	24			13			
	43					19	14
55			11	51	46		15
	21	41	56		17		

The mother is 260. Her four children and four grandchildren are 130. The family is in perfect harmony.

You have been challenged by the legendary strategist, in his letter, to prove yourself as a capable leader of the Shu forces. In order to accomplish this feat, you must be able to:

- Use proper syntax within the R console
- Comment your R code
- Make calculations using formulas
- Distinguish between different types of input and output in the R console

Time is running out. If you aim to prove yourself a worthy leader of the Shu army, then you will need to begin solving Zhuge Liang's puzzle!

Deciphering Zhuge Liang's magic square

Zhuge Liang's puzzle is an 8x8 magic square. In a magic square, all rows, columns, and diagonals add up to the same number. For an 8x8 puzzle like this one, that number is 260. Hence, the *mother* refers to the entire puzzle. Knowing this, take a moment to think about what the *children* and *grandchildren* might refer to.

Continuing, each of the cells in the puzzle hold a number between 1 and 64. Each number appears in one and only one cell. A useful technique for solving a large puzzle is to break it down into smaller components. For example, an 8x8 magic square can be broken down into four 4x4 puzzles (*children*). Furthermore, each 4x4 puzzle can be broken down into four 2x2 squares (*grandchildren*). In this case of *perfect harmony*, each 2x2 and 4x4 puzzle is also a magic square whose number is 130.

Now that we have deciphered Zhuge Liang's puzzle, we can begin solving it with the help of R.

Time for action – solving the first 4x4 magic square

While this puzzle could be solved by hand, it would take a considerable amount of time to do so. Since our deadline is approaching, we will use R to quickly make the necessary calculations.

To simplify our problem, let us first focus on the 4x4 square located in the top-left corner of the puzzle:

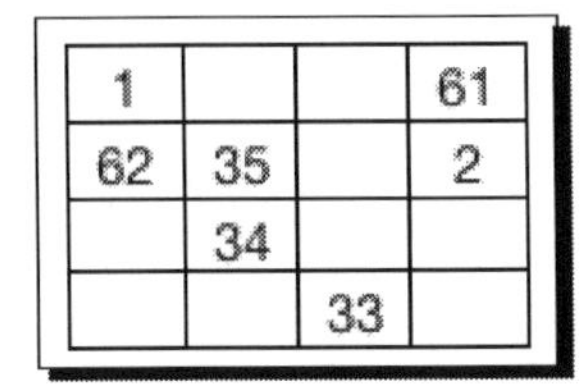

The top-left corner of Zhuge Liang's puzzle

Then follow these simple steps:

1. Open R.
2. On the first line in the R console, next to the greater than sign (>), type the following comment:

```
> #first we will solve the top-left corner of the puzzle
```

3. On the next line in the R console, type the following comment:

```
> #by breaking it down into 2x2 squares and making sure that the
sum of all rows, columns, and diagonals equals 130
```

4. In the upper left-hand corner of the 4x4 square are three numbers. Since we know from Zhuge Liang's note that all 2x2 squares sum to 130, we can calculate the missing value at B1 as follows:

```
> #the value for B1 is:
> 130 - 1 - 62 - 35
```

5. R will display the text `[1] 32`, which indicates that the value of your calculation is 32. Having solved for B1, you can now solve for the missing value in row B, B4. The calculation is as follows:

```
> #the value for B4 is:
> 130 - 32 - 35 - 34
```

6. By working your way through each row, column, and diagonal of the 4x4 puzzle, you can solve the remaining cells in this section, as follows:

```
> #the value for C2 is:
> 130 - 62 - 35 - 2
[1] 31
> #the value for C1 is:
> 130 - 31 - 61 - 2
[1] 36
> #the value for C3 is:
> 130 - 36 - 31 - 33
[1] 30
> #the value for D4 is:
> 130 - 1 - 35 - 30
[1] 64
> #the value for D3 is:
> 130 - 61 - 2 - 64
[1] 3
> #the value for A3 is:
> 130 - 34 - 30 - 3
[1] 63
> #the value for A4 is:
> 130 - 1 - 62 - 63
[1] 4
```

The completed puzzle section is pictured in the following diagram. All calculated cells have been highlighted:

1	32	36	61
62	35	31	2
63	34	30	3
4	29	33	64

Solution to the top-left corner of Zhuge Liang's puzzle

What just happened?

While solving the first quadrant of Zhuge Liang's puzzle, you encountered a number of R's fundamental elements.

Lines

Activity that takes place in the R console is divided into one **line** statements. Long statements will automatically wrap to fit the size of the R window, although they still occur on a single line in the console. For instance, the formulas:

```
> 1 + 1
```

and

```
> 1 + 1 + 2 + 3 + 4 + 5 + 6 + 7 + 8 + 9 + 10 + 11 + 12 + 13 + 14 + 15
+ 16 + 17 + 18 + 19 + 20 + 21 + 22 + 23 + 24 + 25
```

both occupy a single console line, in spite of the fact that the latter formula is wrapped to display on more than one line.

Lines that accept user input begin with a greater than sign (>), whereas static console-generated output lines do not. For example:

```
> date()
[1] "Sun Aug 31 08:00:00 234"
```

The first line accepted user input, whereas the second returned output from the R console.

You can press the *Return* (or *Enter*) key to move from one line to the next, or to commit your code, in the R console. R will always drop down a single line when the *Return* key is pressed. Previous lines will remain displayed in the console, however they will not be editable.

Comments

Each line that begins with a pound sign (#) in the R console is designated as a **comment**. We have used several comments thus far. For example, in the following code:

```
> #the value for A4 is:
> 130 - 1 - 62 - 63
[1] 4
```

The first line is a comment. As is customary in most programming languages, a comment can display any variety of text, code, or other allowable input. Since they are ignored by the console, comments are a useful and necessary tool for documenting and organizing your work.

We inserted several comments into our code in the previous activity. Without them, our console would have been filled with seemingly arbitrary calculations. Instead, our comments provided the context necessary for both others and ourselves to understand what we were calculating and why. It is recommended that you use comments at every relevant opportunity to make your code readable and easy to remember.

Note that no mechanism for multiline comments currently exists in R. However, long comments will automatically wrap to the size of the R console window.

```
> #this is an exceptionally long comment that takes
up the entire width of the R console, so it is
automatically wrapped to display on a second line
```

Alternatively, lengthy explanations can be separated manually by splitting the text into several one line comments, like so:

```
> #this is an exceptionally long comment
> #that has been manually split
> #into several one line comments
```

Note that everything following the pound sign (#) on a given line is ignored by the R console. Therefore, it is possible to combine a comment and other code on the same line, so long as the comment comes last.

Calculations

At its core, R is a sophisticated calculator. We found the value of each missing cell in the 4x4 square using simple mathematical **formulas**. For instance, we used the following formula to find out the value of cell B1:

```
> 130 - 1 - 62 - 35
```

R can, of course, handle an endless variety of calculations. The most commonly used calculations, along with their symbols, are addition (+), subtraction (-), multiplication (*), and division (/). Using R to derive values in this fashion was just our first small step towards becoming familiar and comfortable with the R console.

Output

You may have noticed that some console lines do not begin with the greater than (>) sign. In our preceding activity, these lines contained the results generated by R. The **output** that R returned to our formula for cell B1 is just one example:

```
[1] 32
```

Any time that R displays information to us, it will not be editable and it will not begin with a special prefix. In contrast, all lines that we can edit will begin with the greater than sign (>).

Note also that a `[1]` that appeared before each of our calculated values in the R console output. This is merely R's way of telling us that the result of our formula was contained in a single cell. R likes to think of data in terms of matrices with rows, columns, and cells, and will often prefix its output with such information. You can typically ignore this and only pay attention to the value(s) that you specifically requested.

Visualizing the R console

The following diagram contains a segment of the source code that we created while solving the initial 4x4 puzzle segment. Each comment, calculation, and output has been labeled to demonstrate the visual differences between these types of lines:

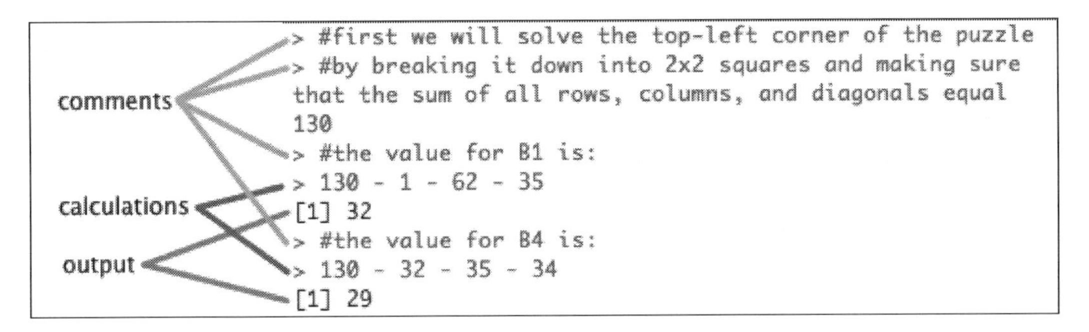
comments
calculations
output
> #first we will solve the top-left corner of the puzzle
> #by breaking it down into 2x2 squares and making sure that the sum of all rows, columns, and diagonals equal 130
> #the value for B1 is:
> 130 - 1 - 62 - 35
[1] 32
> #the value for B4 is:
> 130 - 32 - 35 - 34
[1] 29

Remember that all editable lines begin with a greater than sign (>). Of these, comments begin with a pound sign (#) and are used to document our code. Calculations consist of mathematical operations that we are conducting on our data. Output lines are generated by the R console, are not editable, and lack a leading greater than sign.

Pop quiz

1. Which of the following characters appears at the beginning of each user editable line in the R console?

   a. =

   b. >

   c. -

   d. #

2. Which of the following characters is used to begin a comment line in the R console?
   a. =
   b. >
   c. -
   d. #
3. Which of the following actions submits a line of user input to the R console?
   a. Pressing the *Tab* key
   b. Pressing the *Shift* key
   c. Pressing the *Return* key
   d. Pressing the *Escape* key

Have a go hero

Using the techniques that we employed to solve the top-left quadrant of the puzzle, solve for the remaining cells of Zhuge Liang's 8x8 magic square. You should be able to accomplish this task in a short while as you get accustomed to the R console. Along the way, be sure to identify which console lines are comments, calculations, and outputs. Once finished, check that your row, column, and diagonal values add up to 260. Then, verify your solution with the completed puzzle in the following figure:

1	32	36	61	5	28	40	57
62	35	31	2	58	39	27	6
63	34	30	3	59	38	26	7
4	29	33	64	8	25	37	60
9	24	44	53	13	20	48	49
54	43	23	10	50	47	19	14
55	42	22	11	51	46	18	15
12	21	41	56	16	17	45	52

Solution to Zhuge Liang's 8x8 magic square

Summary

You have now been introduced to the basic operations of R. The R console is composed of lines, which can take the form of comments, calculations, or outputs. You encountered all of these features while solving Zhuge Liang's magic square. In the process, you also gained the skills to:

- Use proper syntax within the R console
- Comment your R code
- Make calculations using formulas
- Distinguish between different types of input and output in the R console

Congratulations, you have earned the late strategist's respect and proven yourself a worthy successor. Zhuge Liang's assistant has provided you with documents containing detailed data about the resources of the Shu and Wei kingdoms. Now it is time for you to organize this information and prepare it for analysis, which will be the focus of *Chapter 4*.

4

Collecting and Organizing Information

After demonstrating your talents by solving Zhuge Liang's puzzle, his assistant provided you with documents summarizing the resources of the Shu army. These documents contain data on gold, equipment, and soldiers. Prior to analyzing these data in R, it is critical that you prepare and organize them. This process will make your subsequent work more clear and efficient.

In this chapter, we will focus on collecting and organizing the information that is available to us. You will encounter several new techniques in R along the way. By the end of this chapter, you will be able to:

- Import external data into R
- Use variables to organize and manipulate your data
- Manage the R workspace

Time for action – importing external data

Our first task is to pull external resource data into R, so we can begin to examine it. To accomplish this, open the R console and proceed through the following steps:

1. Set your R working directory using the `setwd(dir)` function. The path used in the following code acts as an example. Your working directory should be set to a relevant location on your own computer:

```
> #set the R working directory
> #replace the sample location with one that is relevant to you
> setwd("/Users/johnmquick/rBeginnersGuide/")
```

2. Copy the `hanzhongResources.csv` file into your R working directory. This file contains resource information for the Shu forces that are currently recuperating in Hanzhong.

3. Read the resource file into R using the `read.csv(file)` command:

```
> #use read.csv(file) to read an external data file into R
> #Shu resources located in Hanzhong, China
> read.csv("hanzhongResources.csv")
```

4. R will read and display the contents of the file, and the result is shown in the following screenshot:

```
> hanzhongResources
     Gold Provisions Soldiers EquipmentCondition
1 1000000    1000000   100000                  1
```

These data indicate that your forces in Hanzhong currently have 1,000,000 each of gold and provisions, 100,000 soldiers, and equipment that is in mint condition.

What just happened?

After setting your working directory, you encountered a new function. Its syntax differs from the commands that we have previously used.

read.csv(file)

In `read.csv(file)`, a period is placed between the function name `read` and the `csv` attribute. The term `csv` told the `read` function that the data in our file contained comma-separated values. It is important to distinguish which `read` function we want to use, because it can take on a number of alternative forms, such as `read.S` and `read.SPSS`.

The `file` portion of the `read.csv(file)` function is similar to `dir` in `setwd(dir)`. Since we placed our data file in our working directory, the `file` argument needed only to contain a file name and extension. Had the data been placed elsewhere, a complete file path would have been necessary.

comma-separated values (csv) files

Throughout this book, we will use **comma-separated values**, or **CSV**, data files. This is the recommended file type for importing data into R. However, you should be aware that R can accept data from a wide variety of sources. Therefore, you can typically import from whichever sources you may use.

Pop quiz

1. What is the key difference between the function arguments `dir` and `file`?
   a. The `dir` argument contains a path, whereas the `file` argument contains a filename.
   b. The `dir` argument contains a path to a directory folder, whereas the `file` argument contains a path to a file.
   c. Functions beginning with `read` receive the `file` argument, whereas functions beginning with `set` receive the `dir` argument.
   d. There is no difference between the `dir` and `file` arguments.

Time for action – creating and calling variables

Although reading your data into R allows you to visualize it in the console and use it to make hand-typed calculations (as we did in *Chapter 3*), you generally need a more organized and flexible method for manipulating your data. R variables are well suited to accomplish this aim. Instead of only reading our resource data into R, let us this time read and store our data in a variable:

1. Use the following code to store the data from our resource file in a variable named `hanzhongResources`:

```
> #read the data from hanzhongResources.csv into a variable
named hanzhongResources
> hanzhongResources <- read.csv("hanzhongResources.csv")
```

2. Note that R did not display any output after step 1 and simply dropped down to the next line in the console. To verify the contents of our new variable, we must call it by typing its name in the R console.

```
> #display the contents of the hanzhongResources variable
> #Shu resources located in Hanzhong, China
> hanzhongResources
```

3. R will display the contents of the variable.

```
> hanzhongResources
     Gold Provisions Soldiers EquipmentCondition
1 1000000    1000000   100000                  1
```

You may have noticed that calling your `hanzhongResources` variable yields the exact same output as reading the original CSV file into R. However, the variable is much more efficient, because we do not have to type the entire `read.csv(dir)` code each time we want to display its data. Instead, we may simply refer to it by name.

What just happened?

You have created and called your first **variable** in R. Variables are essential for storing and manipulating data. Each time you create a variable in R, you will follow a similar process to the one we just exercised. The four steps in the variable creation process are as follows:

1. **Start with the variable name**

   In our preceding example, `hanzhongResources` was the name of our variable. A name should be the first thing that appears on a new console line when creating an R variable.

2. **Add less than minus (`<-`)**

   After the variable name, the less than minus symbol, or `<-`, should be added. The `<-` symbol can be thought of as meaning "is set equal to the contents of." These characters have the effect of assigning the information on their right to the variable name on their left. For example, the line `> A <- B` can be read as "the variable named `A` is set equal to the contents of `B`." Therefore, in our previous example, we set the variable named `hanzhongResources` equal to the contents of the file `hanzhongResources.csv`.

3. **Add the data source**

   The data source `hanzhongResources.csv` was used in our example. A data source should be the last thing that appears on the console line when creating an R variable. Data sources typically take on the form of datasets that are read into R, numeric values, or previously created variables.

4. **Verify the variable's contents**

   When executing a line of R code does not yield visible output, as is the case when creating a new variable, it is wise to verify the results of our actions. To display the contents of a variable, type its name in the R console and press the *Return* key. In our case, entering `hanzhongResources` will yield a console output containing the Shu army's resources located in Hanzhong, China.

Pop quiz

1. Which of the following is not an advantage of storing the output of the `read.csv(file)` function as a variable?

   a. The variable name is more efficient to type.

   b. The variable name is easier to remember.

   c. The variable's data is preserved even if the original CSV file is moved or deleted.

   d. The variable explicitly states its data source.

2. Interpret the following R console line in words:

```
> myVariable <- myData
```

   a. The variable `myVariable` is set equal to the contents of `myData`.
   b. The variable `myData` is set equal to the contents of `myVariable`.
   c. The variable `myVariable` is less than negative `myData`.
   d. The variable `myVariable` is greater than zero and less than negative `myData`.

Have a go hero

You are now familiar with the process behind creating a new data variable in R. The `soldiersByCity.csv` file contains the total number of soldiers located in each major city within Shu and Wei territory. Copy this file into your R working directory. Then use the four step process to create and verify the contents of a new variable called `soldiersByCity`. This variable should contain all of the data in the `soldiersByCity.csv` file.

Time for action – accessing data within variables

Both our `hanzhongResources` and `soldiersByCity` variables contain a complete set of values (as opposed to a single value). We already know that typing a variable's name into R will output all of its contents in the console. However, we often need to access the columns, rows, and cells within a dataset to perform calculations.

We will start by exploring two methods for accessing the columns in our `soldiersByCity` variable:

1. First, we will access the `Soldiers` column from our `soldiersByCity` variable through R's `variable$column` notation:

```
> #isolate a single column within a dataset using the
variable$column notation.
> #display the contents of the Soldiers column from the
soldiersByCity variable
> soldiersByCity$Soldiers
```

2. R will display the contents of the **Soldiers** column, and the result is shown in the following screenshot:

```
> soldiersByCity$Soldiers
 [1] 100000  25000  15000  15000  10000  10000
 [7]   5000   5000   5000   5000   5000 200000
[13]  50000  25000  20000  15000  15000  15000
[19]  15000  15000  15000  10000
```

3. This time, let us use the `attach(variable)` function to simplify our operation.

```
> #isolate a single column within a dataset using the
attach(variable) function and simplified notation
> #attach the soldiersByCity variable
> attach(soldiersByCity)
> #display the contents of the Soldiers column from the
soldiersByCity variable
> Soldiers
```

4. R will display the contents of the `Soldiers` column:

```
> Soldiers
 [1] 100000  25000  15000  15000  10000  10000
 [7]   5000   5000   5000   5000   5000 200000
[13]  50000  25000  20000  15000  15000  15000
[19]  15000  15000  15000  10000
```

Next, we will access a single row within the `soldiersByCity` variable:

5. Use the `variable[row, column]` matrix notation to display the contents of the tenth row in our `soldiersByCity` variable:

```
> #isolate a single row within a dataset using the
variable[row, column] matrix notation.
> #display the contents of the tenth row in the soldiersByCity
variable
> soldiersByCity[10,]
```

6. R will display the contents of the tenth row in our `soldiersByCity` dataset:

```
> soldiersByCity[10,]
   Kingdom    City Soldiers
10     Shu Yunnan     5000
```

7. Similarly, we can use matrix notation to access a single cell within our dataset.

   Use matrix notation to display the contents of cell `[5,3]` in our `soldiersByCity` variable:

```
> #isolate a single cell within a dataset using the
variable[row, column] matrix notation.
> #display the contents of cell [5,3] in the soldiersByCity
variable
> soldiersByCity[5,3]
```

8. R will display the contents of cell `[5,3]`, as shown:

```
> soldiersByCity[5,3]
[1] 10000
```

What just happened?

You have just practiced accessing data within a variable from each possible angle, that is, by columns, rows, and individual cells. Let us take a closer look at how variable data is accessed in R.

variable$column notation

Individual columns within a dataset can be accessed via the **variable$column** notation. Think of the dollar sign (`$`) as the letter S, as in the word "select." In this way, the notation can be read in words. For example, the line `> A$B` can be read as "from variable A, select column B." During our activity, we selected the `Soldiers` column from the `soldiersByCity` variable by typing the following code in the R console:

```
> soldiersByCity$Soldiers
```

attach(variable) function

The `attach(variable)` function is a convenient way to relieve ourselves of lengthy notation in some, but not all, cases. When a variable is **attached** in the R console, its columns can be referred to by name, without the need to identify the variable. For example, after we attached `soldiersByCity`, we could display the contents of the `Soldiers` column by simply typing `> Soldiers` in the console.

A caveat with the `attach(variable)` function is that often only a single variable can be attached to the R console at a given time. For instance, if we were to attach both our `hanzhongResources` and `soldiersByCity` variables at the same time, we would run into a problem regarding the `Soldiers` column. Since both of these variables contain such a column, R can only refer to the most recently attached version. Accessing the other would require the use of `variable$column` notation. In fact, R will warn you if you attach two variables that share a common column name. The following error occurs when the `soldiersByCity` variable is attached, followed by `hanzhongResources`:

```
The following object(s) are masked from soldiersByCity :

 Soldiers
```

On the other hand, attaching a variable can be useful and efficient when you are working with a single, large dataset. If you are only manipulating data from one variable, then you will not run into the demonstrated error. Furthermore, you can always have one variable attached, even if you are working with datasets that have identical column names. Of course, if your variables do not have columns in common, then attaching them all is an option. In any case, you can always refer to columns using `variable$column` notation, which we will do throughout the remainder of this book.

Note that should you ever need to **detach** a variable, you can use the `detach(variable)` function. This will return the variable to its prior status in the console, as if it had never been attached in the first place.

variable[row, column] notation

When referring to row data or individual cells, the **variable[row, column]** notation should be used. For rows, such as when we accessed the tenth row in `soldiersByCity` via `> soldiersByCity[10,]` the column portion of the notation is omitted. This tells R to retrieve all of the columns in the row.

To isolate an individual cell, both a row and column value must be specified. When we accessed cell `[5,2]` from `soldiersByCity` via `> soldiersByCity[5,2]` the `5` represented the cell's row, whereas the `2` defined the cell's column. This is similar to selecting a single point from a graph using its x-y coordinates, except the graph in our case is a matrix of data values.

On a side note, you may have noticed that `variable[row,column]` notation can also be used to refer to columns. This can be accomplished by leaving the row portion of the notation blank. For example, to access the `City` column in `soldiersByCity`, we could use the code `soldiersByCity[,1]`, this tells R to retrieve every row within the `City` column.

Pop quiz

1. Interpret the following R console line in words:

   `> myVariable$myColumn`

   a. Multiply the data within `myVariable` by the data within `myColumn`.

   b. Divide the data within `myVariable` by the data within `myColumn`.

   c. In variable `myColumn`, select column `myVariable`.

   d. In variable `myVariable`, select column `myColumn`.

2. Under which of the following circumstances is it best not to attach dataset variables in the R console?
   a. You are working with a single dataset.
   b. You are working with multiple datasets that contain identical column names.
   c. You are working with multiple datasets that contain identical column names, but want to attach only one of them.
   d. You are working with multiple datasets that do not contain identical column names.

3. The `variable[row,column]` notation can be used to access data from which of the following locations?
   a. Rows.
   b. Columns.
   c. Cells.
   d. All of the above.

Time for action – manipulating variable data

Being able to access the information stored in a variable is the initial step towards manipulating its data. Variables and their data can be used in the same way that we used numbers to perform calculations in *Chapter 2*. They can be used in mathematical formulas as well as in function arguments.

1. Use your `hanzhongResources` variable to calculate the amount of resources that the Shu army would have remaining if a flood were to destroy 75% of each resource:

```
> #if a flood destroyed 75% of the Shu resources at Hanzhong,
how much of each resource would remain?
> #multiply the hanzhongResources variable by 0.25 to represent
the remaining 25% of the original resources
> hanzhongResources * 0.25
```

2. R will display the result of the calculation:

```
> hanzhongResources * 0.25
    Gold Provisions Soldiers EquipmentCondition
1 250000     250000    25000               0.25
```

3. Now assume that the hypothetical flood only affected the provisions at Hanzhong, while all of the other resources remained unharmed. Here, you must perform a calculation only on the `Provisions` column of the `hanzhongResources` variable:

```
> #if a flood destroyed 75% of the Provisions at Hanzhong,
how much would remain?
> #multiply the Provisions column by 0.25 to represent the
remaining 25% of the original resources
> hanzhongResources$Provisions * 0.25
```

4. R will display the results of the calculation. Note that calculations can be applied in the same fashion to rows, columns, and cells.

```
> hanzhongResources$Provisions * 0.25
[1] 250000
```

5. Variable data can also be used in function arguments. On a less disastrous note, use your `soldiersByCity` variable to calculate the mean (average) number of soldiers stationed in a Shu city:

```
> #use the mean(data) function to calculate the average number
of soldiers stationed in a Shu city
> #on average, a Shu city has this many soldiers:
> mean(soldiersByCity$Soldiers)
```

6. R will display the results of the calculation. Note that functions can be applied in the same fashion to row, column, or cell data, or entire datasets.

> mean(soldiersByCity$Soldiers)
[1] 27045.45

7. Moreover, calculation results can be saved into new variables for use at a later time. This time, save the calculation from step 5 into a new variable named `meanSoldiersByCity`:

```
> #save the mean number of soldiers per city into a new
variable named meanSoldiersByCity
> meanSoldiersByCity <- mean(soldiersByCity$Soldiers)
```

8. R will not return any output. Verify the contents of `meanSoldiersByCity` by entering it into the R console:

```
> #display the contents of meanSoldiersByCity
> meanSoldiersByCity
```

9. R will display the contents of the `meanSoldiersByCity` variable:

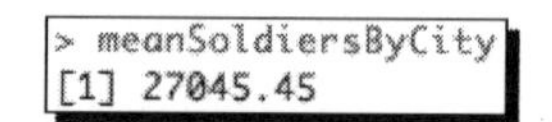
> meanSoldiersByCity
[1] 27045.45

What just happened?

In just a few lines of code, you have experienced the range of variable manipulations that you will use on a regular basis in R. Let us explore each one individually.

Performing a calculation on an entire dataset

When you used your `hanzhongResources` variable to calculate the consequences of a flood across each resource, you discovered that when a variable is manipulated in this manner, so is all of its underlying data.

For demonstration, consider the following table with the cell values of 1, 2, 3, and 4 in columns a, b, c, and d respectively:

a	b	c	d
1	2	3	4

Suppose that this table is saved in a R variable named `lettersAndNumbers`. If we were to add one to the `lettersAndNumbers` variable in R, by the following command:

```
> lettersAndNumbers + 1
```

Our resulting table would contain the addition of each cell's value and one, as follows:

a	b	c	d
2	3	4	5

As you can see, R will attempt to perform any calculation made on a dataset to each of its values. However, it is worth noting that R will not always be able to make a successful calculation on every cell in a dataset.

For instance, if we tried to make a numeric calculation on the `Kingdom` and `City` columns of our `soldiersByCity` variable, R would return a warning along with an `NA` or *not applicable* values. This is due to the fact that our `Kingdom` and `City` columns contain text and therefore it does not make sense to manipulate them numerically. To see this warning in action, enter the following lines into the R console:

```
> #what happens if we try to make a numeric calculation on
nonnumeric data?
> #we receive a warning, because it does not make sense to
manipulate text mathematically
> soldiersByCity * 5
```

This would result in the following screen:

```
> soldiersByCity * 5
   Kingdom City Soldiers
1       NA   NA   500000
2       NA   NA   125000
3       NA   NA    75000
4       NA   NA    75000
5       NA   NA    50000
6       NA   NA    50000
7       NA   NA    25000
8       NA   NA    25000
9       NA   NA    25000
10      NA   NA    25000
11      NA   NA    25000
12      NA   NA  1000000
13      NA   NA   250000
14      NA   NA   125000
15      NA   NA   100000
16      NA   NA    75000
17      NA   NA    75000
18      NA   NA    75000
19      NA   NA    75000
20      NA   NA    75000
21      NA   NA    75000
22      NA   NA    50000
Warning messages:
1: In Ops.factor(left, right) : * not meaningful for factors
2: In Ops.factor(left, right) : * not meaningful for factors
```

Here, the `Soldiers` columns contain numeric values and therefore each value within it is successfully multiplied by five. However, the text in the `Kingdom` and `City` columns cannot be multiplied. Hence, a warning message is returned. To avoid deriving meaningless values and upsetting the R console, it is important to be aware of your data and apply appropriate calculations to them.

Performing a calculation on a row, column, or cell

Manipulating row, column, or cell data is identical to manipulating an entire dataset contained within a variable. The difference is not in the calculation, but rather in what you choose to perform the calculation on. Depending on whether you aim to manipulate row, column, or cell data, you will need to access the values in the appropriate manner. See the *Accessing data within variables* section of this chapter for a review of these methods.

Using variable data in function arguments

A variable's data, be it from the entire set or a specific subset (row, column, or cell), can be used in function arguments. Our preceding activity used the `mean(data)` function to calculate the average number of soldiers among the Shu cities listed in our `soldiersByCity` variable. We could have easily done the same with the entire `soldiersByCity` dataset, a single row, or an individual cell. The best method for using variable data in arguments will depend on the goal of the manipulation and the specific function being employed.

Saving a variable calculation into a new variable

Do not forget that a variable's purpose is to store and organize your information. Quite often, we will need to store the results of a calculation or function into a new variable for subsequent manipulation. The body of variables and other objects that we amass throughout our work are stored in the R **workspace**, which is the topic of our next section.

Pop quiz

The table `myTable` contains two rows, three columns, and six cells with the numbers one through six. Use this table to answer questions 1 and 2.

myTable		
1	2	3
4	5	6

1. Consider the following line of code:

   ```
   > myTable * 10
   ```

   If this code were applied to `myTable`, what would be the result? Write the appropriate values in the blank cells of `myTableAfterManipulation1`:

myTableAfterManipulation1

2. Consider the following line of code:

   ```
   > myTable[1,2] + 10
   ```

   If this code were applied to `myTable`, what would be the result? Write the appropriate values in the blank cells of `myTableAfterManipulation2`:

myTableAfterManipulation2

3. Interpret the following R console line in words:

```
> myVariable <- mean(myData$myColumn)
```

   a. Calculate the mean of `myColumn` and then set `myVariable` equal to the result.
   b. Calculate the mean of `myData` and then set `myVariable` equal to the result.
   c. In `myData`, select `myColumn`, calculate its mean, and then set `myVariable` equal to the result.
   d. Set `myVariable` equal to the contents of `myData` and then calculate its mean.

Have a go hero

To practice the variety of methods that we have covered for manipulating variables, use your resource data and knowledge of R to complete the following tasks:

1. Suppose you are concerned with the potential of flooding to damage your resources. Calculate the amount of resources that would remain if a flood destroyed half of each resource stored in your `hanzhongResources` variable. Save the results into a single variable named `hanzhongResourcesAfterFlood`.
2. To account for a recent relocation of 5000 soldiers from Guanghan to Baxi, subtract 5000 from the cell representing the number of Guanghan soldiers and add 5000 to the cell representing the number of Baxi soldiers in the `soldiersByCity` variable. Save each of these calculations into a new variable. The variables should be named `guanghanSoldiersAfterRelocation` and `baxiSoldiersAfterRelocation` respectively.
3. Use the `min(data)` and `max(data)` functions and your `soldiersByCity` variable to calculate minimum and maximum number of soldiers in either army by city. Save the results as variables named `minSoldiersByCity` and `maxSoldiersByCity` respectively.
4. Use the `sum(data)` function and your `soldiersByCity` variable to calculate the total number of soldiers in the Shu and Wei armies. Then, save the result as a variable named `totalSoldiers`.

If you encounter a warning or error during any of these tasks, think about how you can be more specific about which data you want to apply your calculation or function to. For detailed information on handling these occurrences, refer back to the *Performing a calculation on an entire dataset* section of this chapter.

Time for action – managing the R workspace

The R workspace stores all user-generated objects (variables in our case) that are created during a session. Its contents can be saved and loaded for future use.

1. Use the `ls()` function to display a list of your R workspace contents:

```
> #list the current contents of the R workspace
> ls()
```

2. R will display a list of the objects in your workspace:

```
> ls()
[1] "baxiSoldiersAfterRelocation"
[2] "guanghanSoldiersAfterRelocation"
[3] "hanzhongResources"
[4] "hanzhongResourcesAfterFlood"
[5] "maxSoldiersByCity"
[6] "meanSoldiersByCity"
[7] "minSoldiersByCity"
[8] "soldiersByCity"
[9] "totalSoldiers"
```

3. Use the `save.image(file)` function to save your R workspace to your working directory. The `file` argument should contain a meaningful filename and the `.RData` extension:

```
> #save the R workspace to the working directory using
save.image(file)
> save.image("rBeginnersGuide_Ch_04.RData")
```

4. R will save your workspace file. Browse to the working directory on your hard drive to verify that this file has been created.

5. Use the `q()` command to quit R. Ignore or decline any messages that you receive.

```
> #quit an R session
> q()
```

6. R will close.

7. Relaunch R by double-clicking on its icon. Then use the `ls()` command to verify that the current workspace is empty:

```
> #list the current contents of the R workspace
> ls()
```

You will be presented with the following result:

8. Use the `load(file)` function to load your saved workspace file. The `file` argument should be identical to what you used in step 3:

```
> #load a previously saved R workspace using load(file)
> load("rBeginnersGuide_Ch_04.RData")
```

9. Use the `ls()` command to verify that the saved contents are now contained in the R workspace:

```
> #list the current contents of the R workspace
> ls()
```

```
> ls()
[1] "baxiSoldiersAfterRelocation"
[2] "guanghanSoldiersAfterRelocation"
[3] "hanzhongResources"
[4] "hanzhongResourcesAfterFlood"
[5] "maxSoldiersByCity"
[6] "meanSoldiersByCity"
[7] "minSoldiersByCity"
[8] "soldiersByCity"
[9] "totalSoldiers"
```

What just happened?

You just exercised the primary workspace management functions that you will need to carry your work through multiple R sessions. These included listing, saving, and loading the contents of your R workspace.

Listing the contents of the R workspace

The R workspace contains every object that you have created during an R session. Up to this point, our objects have taken the form of variables that either read data from CSV files or store the results of calculations. The `ls()` function can be called at any time to list the contents of your R workspace.

Saving the contents of the R workspace

To save the R workspace, use the `save.image(file)` function. Since we were operating within our R working directory, the `file` argument needed only to contain our desired filename and the `.Rdata` extension. Alternatively, if you were to save your workspace to a different location on your hard drive, then you would need to enter a complete path in the `file` argument. Be sure to always include the `.Rdata` extension when saving your workspace, as it is necessary for R to be able to recognize the file when loaded.

Loading the contents of the R workspace

To load an R workspace file, use the `load(file)` function. Here, the `file` argument is identical to the one received by the `save.image(file)` function. Hence, if the file you want to load is contained within your working directory, you need only to use the file name and `.RData` extension. If it is housed elsewhere, then you will need to use a complete file path.

Note that, depending on your version of R, saving and loading of the R workspace can be automated on launch or quit, or accomplished by clicking through the menu options. You may want to explore the available menu choices and preference settings available to you. This will let you configure R to best suit your workflow. Nevertheless, it is recommended that you continue to use the R console to manage your workspace, because it gives you complete control over your work.

Quitting R

As you have witnessed, the `q()` command can be used to exit R. You can, of course, use menu options, keyboard commands, or other methods available on your computer to quit R.

Unless you have specifically told R to save your workspace on exit, all of its contents will be lost. Remember to save your workspace before quitting R.

Distinguishing between the R console and workspace

When you relaunched R and loaded your saved workspace file in the preceding activity, you may have noticed that the contents of your R console were not retrieved. This reveals an important distinction between the R console and the R workspace. Essentially, the workspace stores all of your objects, whereas the console contains a log of everything that has been done to and with those objects.

Consider the act of watching a movie in a theater as an analogy to demonstrate the relationship between the workspace and console. The audience members, movie screen, and chairs are all located within the same room (the workspace). Everything that these entities do—sneeze, laugh, chomp, display the movie, get chewing gum stuck to them—is recorded in the history of the movie experience (the console).

Thus, the workspace contains objects (such as the people in a movie theater) and the console logs the interactions between them (such as one patron spilling popcorn on the head of another).

Saving the R console

Since the console is not saved along with the workspace, you may be wondering how you can preserve the information logged in the R console during a session. While there is no function available in R that allows us to save its contents, we do have other options.

One is to copy and paste the R console into a text editor. Another, depending on your version of R, may be to use the menu to save a copy of your console as a text file. These are the preferred techniques for capturing the R console, although you may be able to think of alternative methods.

In any event, it is highly recommended that you save the R console at the end of every session. Having the log of your previous work can be invaluable to the prevention of rework and to informing your future work. It can also help you organize and remember everything about your current project, especially when you have a large amount of data and many objects to manage.

Pop quiz

1. When saving the R workspace, which of the following extensions should you include?
   a. `.txt`
   b. `.R`
   c. `.RData`
   d. No extension is necessary

2. Which of the following best describes the relationship between the R console and the R workspace?
   a. The R workspace and R console can both be saved using the `save.image(file)` function.
   b. The contents of the R workspace and R console can both be displayed using the `ls()` function.
   c. The R console stores objects, whereas the R workspace logs the actions related to those objects.
   d. The R workspace stores objects, whereas the R console logs the actions related to those objects.
3. Which of the following is not an option for saving the R console?
   a. Using a built-in R function.
   b. Copying the console contents into a text editor.
   c. Using the R menu options to save the console as a text file.
   d. Taking a screenshot of the R console.

Have a go hero

Your final challenge for this chapter will be to collect and organize the remaining portions of Zhuge Liang's resource data. This will entail reading CSV data into R, creating new variables, accessing and manipulating variable data, and saving your R workspace and console. Demonstrate your mastery of these concepts by preparing historic battle data for analysis through the following actions:

1. The `battleHistory.csv` file contains data from 120 previous battles between the Shu and Wei forces. Read these data into an R variable named `battleHistory`.
2. Use the data imported in step 1 to answer the following question. What is the average number of soldiers to engage in combat for both the Shu and Wei forces? Save your results into separate variables, named `meanSoldiersShu` and `meanSoldiersWei` respectively.
3. Save the contents of your R workspace into a new file named `rBeginnersGuide_ch_04_hero.RData`.
4. Save the contents of your R console into a new text file named `rBeginnersGuide_ch_04_hero.txt`.

Feel free to refer back to the previous sections in this chapter for assistance with collecting and organizing this information.

Summary

Throughout this chapter, you deeply explored methods for collecting and organizing your data in R. These techniques are critical to your success as a strategist and an analyst. Being able to manage your data efficiently and effectively is key to simplifying your workflow and making your results intelligible to others. While collecting and organizing your data, you acquired the skills necessary to:

- Import external data into R
- Use variables to organize and manipulate your data
- Manage your R workspace

Now that your data are prepared, you can begin to assess the military challenges that await the Shu forces. In *Chapter 5*, we will weigh the potential combat options available to the Shu army. It is up to you to set forth a prudent course of action.

5

Assessing the Situation

With our data prepared for analysis, we can now consider the potential combat options available to the Shu army. Ultimately, you have the responsibility to use these data to make practical and meaningful decisions about future courses of action. To achieve success, you will need to fully consider the situation and form a sound basis for reasoned decision-making. This requires you to build upon the techniques that we practiced in Chapter 4 and to explore new ones in R.

In this chapter, we will focus on assessing the information that is available to us and using it to weigh potential decisions. By the end of this chapter, you will be able to:

- Use multi-argument and variable-argument functions to perform calculations
- Create predictive models using regression analysis
- Consider the statistical and practical significance of your analyses

Time for action – making an initial inference from our data

In *Chapter 4*, we saved our R workspace for the first time. As you saw in the previous chapter, we can use the `load(file)` function to continue where we left off.

You also created variables to hold the mean number of soldiers engaged for the Shu and Wei forces, based on historical data from 120 battles between the kingdoms. Let us make an initial inference about these values:

1. Open R and set your working directory, as follows:

```
> #set the R working directory
> #replace the sample location with one that is relevant to you
> setwd("/Users/johnmquick/rBeginnersGuide/")
```

2. Load the *Chapter 5* workspace. It contains the information that we generated in *Chapter 4* and will continue to work on in this chapter:

```
> #load the chapter 5 workspace
> load("rBeginnersGuide_Ch_05_ReadersCopy.RData")
> #verify the contents of the workspace
>ls()
```

3. Display the mean number of Shu and Wei soldiers engaged in past battles. We saved these values into variables in the previous chapter.

```
> #mean number of Shu soldiers engaged in battle
> meanSoldiersShu
[1] 21035.83
> #mean number of Wei soldiers engaged in battle
> meanSoldiersWei
[1] 21937.5
```

4. Calculate the ratio of mean Wei soldiers to Shu soldiers and save it to a new variable named `meanSoldierRatioWeiShu`. Then display the result:

```
> #ratio of mean Wei soldiers to Shu soldiers
> meanSoldierRatioWeiShu <- meanSoldiersWei / meanSoldiersShu
> #display the contents of meanSoldierRatioWeiShu
> meanSoldierRatioWeiShu
[1] 1.042863
```

5. Predict the number of Wei soldiers that would engage in combat if the Shu prepared 100,000 soldiers for battle:

```
> #how many Wei soldiers would we expect to engage in battle if
our Shu forces numbered 100,000?
> 100000 * meanSoldierRatioWeiShu
[1] 104286.3
```

What just happened?

After preparing R, we used our historic battle data to calculate the ratio of the mean Wei soldiers engaged in past conflicts to the mean number of Shu soldiers. The ratio value of 1.04 suggests that the Wei army brings roughly 4% more soldiers into battle than Shu does on average. We can use this ratio in our predictions and plans for future battles. In general, we expect the Shu army to be outnumbered regardless of the conflict. Inferences like this one may have implications for the combat strategies that we choose to employ.

Examining our data

Before we move into deeper analyses, let us take a moment to examine our battle history data. This will help us better understand the information that we are working with. Display the contents of your `battleHistory` variable by entering its name into the R console:

```
> #display all of our battle history data
> battleHistory
```

```
> battleHistory
      Method Rating SuccessfullyExecuted  Result ShuSoldiersEngaged WeiSoldiersEngaged DurationInDays
1  headToHead     15                    Y  Defeat               5000              15000            116
2  headToHead     25                    Y  Defeat               5000              10000             96
3  headToHead     50                    Y  Defeat               5000              10000             76
4  headToHead     55                    Y  Defeat               5000              10000             61
5  headToHead     60                    Y  Defeat               7500              15000             52
6  headToHead     25                    Y  Defeat              50000             100000             94
7  headToHead     30                    Y  Defeat             100000             200000             87
8  headToHead     90                    Y Victory              10000               5000             36
9  headToHead     90                    Y Victory             100000              50000             44
10 headToHead     80                    Y Victory              30000              15000             49
```

Note that only the first 10 of the total 120 rows are shown here

As you can see, our dataset is composed of seven columns, each containing valuable information about past battles between the Shu and Wei forces:

- `Method`: contains the type of battle technique employed. These are `headToHead`, `surround`, `ambush`, and `fire`.
- `Rating`: contains a measure of the Shu army's performance on a scale from `0` to `100`. After each battle, Zhuge Liang rated the Shu army to keep a record of its performance under diverse combat conditions.
- `SuccessfullyExecuted`: contains a yes (`Y`) or no (`N`) value indicating whether the battle method was executed successfully.
- `Result`: tells us whether the battle ended in `Victory` or `Defeat`.
- `ShuSoldiersEngaged`: presents the number of soldiers who engaged in combat for the Shu army during each battle.
- `WeiSoldiersEngaged`: is identical to `ShuSoldiersEngaged`, but for the Wei forces.
- `DurationInDays`: indicates how long each battle lasted, in days.

Overall, data from 120 battles are included, with each combat method represented 30 times. Now that we are more aware of our data, let us begin analyzing it in more detail.

Time for action – creating a subset from a large dataset

We will start by assessing the feasibility of head to head combat with the Wei army. Since we have past data related directly to head to head battles, we should specifically target this information in order to best address the method's prospects. Currently, those data are part of a large set that also contains information on other methods. However, we can use the multi-argument function `subset(data, ...)` to isolate our head to head combat data and simplify our analysis of this strategy:

1. Create a subset of data using the `subset(data, ...)` function and save it to a new variable named `subsetHeadToHead`:

```
> #use the subset(data, ...) function to create a subset from a
larger dataset
> #create a subset that isolates our head to head combat data
> subsetHeadToHead <- subset(battleHistory, battleHistory$Method
== "headToHead")
```

2. Verify the contents of the new subset. Note that the console should return thirty rows, all of which contain `headTohead` in the `Method` column:

```
> #display the contents of the head to head subset
> subsetHeadToHead
```

```
> subsetHeadToHead
       Method Rating SuccessfullyExecuted  Result ShuSoldiersEngaged WeiSoldiersEngaged DurationInDays
1  headToHead      5                    Y  Defeat               5000              15000            116
2  headToHead     15                    Y  Defeat               5000              10000             96
3  headToHead     25                    Y  Defeat               5000              10000             76
4  headToHead     25                    Y  Defeat               5000              10000             61
5  headToHead     35                    Y  Defeat               7500              15000             52
6  headToHead     20                    Y  Defeat              50000             100000             94
7  headToHead     30                    Y  Defeat             100000             200000             87
8  headToHead     85                    Y Victory              10000               5000             36
9  headToHead     95                    Y Victory             100000              50000             44
10 headToHead     85                    Y Victory              30000              15000             49
11 headToHead     10                    Y  Defeat               2500               5000            112
12 headToHead     15                    Y  Defeat               2000               2500             90
13 headToHead     15                    Y  Defeat                250                500            111
14 headToHead     20                    Y  Defeat               1000               2000             93
15 headToHead     10                    Y  Defeat               7000               7500            120
16 headToHead     10                    Y  Defeat               5000               7500            100
17 headToHead     90                    Y Victory              15000              10000             35
18 headToHead     80                    Y Victory              15000              10000             45
19 headToHead     85                    Y Victory              25000              10000             40
20 headToHead     85                    Y Victory              25000              20000             45
21 headToHead     35                    Y  Defeat              30000              35000             95
22 headToHead     45                    Y  Defeat              25000              35000            105
23 headToHead     50                    Y  Defeat              40000              45000            100
24 headToHead     35                    Y  Defeat              30000              45000             91
25 headToHead     25                    Y  Defeat              65000              75000            120
26 headToHead     20                    Y  Defeat              50000              75000             99
27 headToHead     30                    Y  Defeat              60000              75000            102
28 headToHead     90                    Y Victory              75000              40000             44
29 headToHead     90                    Y Victory              50000              25000             50
30 headToHead     95                    Y Victory             100000              60000             30
```

What just happened?

In the one console line that it took to create a subset of our data, you encountered your first **multi-argument** (and **variable-argument**) function in the R language.

Multi-argument functions

You were first introduced to functions in *Chapter 2*. There, the `date()` function received no arguments and output the current date and time in the R console. Shortly after, you used `setwd(dir)` and `getwd(dir)` to set and retrieve your R working directory. Both of these functions received a single argument. With `subset(data, ...)` you have used your first multi-argument function. Further, `subset(data, ...)` represents a variable-argument function, meaning that the exact number of arguments it receives can be different depending the circumstance. In our example, we used two arguments. However, we could have used more arguments to further specify our subset. For instance, we could have added an additional argument to our `subset(data, ...)` function that told R to include only certain columns in its output.

Variable-argument functions

Any time that you see ellipsis (`...`) in an R function, you know that it accepts a variable number of arguments. In contrast, some multi-argument functions, such as `cor(x, y, use, method)` for correlations, accept no more and no less than a specific number of arguments. However many others, such as `plot(x, y, ...)` for scatterplots, can accept relatively few or many arguments, depending on the situation.

Equivalency operators

In the second argument of our `subset(data, ...)` function, we employed the **equivalency operator**. It is formed by two consecutive equals signs (==). This operator evaluates the equivalency of two statements, the one to its left and the one to its right. If the statements are equal, then the argument is deemed `True`. If not, it is considered `False`.

Conversely, the **non-equivalency operator**, which is formed by an exclamation point joined with a single equals sign (`!=`), tests to see if two statements are not equal. If they indeed are not, then the argument is deemed `True`, otherwise `False`.

subset(data, ...)

Our implementation of the `subset(data, ...)` function made use of two arguments. The first referred to our data source, the `battleHistory` variable. The second specified the exact data that we wanted to pull from that source.

```
> subsetHeadToHead <- subset(battleHistory, battleHistory$Method
== "headToHead")
```

In our case, we wanted to include battles only if they employed the head to head combat method. To clarify this operation, let us dissect the second argument.

```
battleHistory$Method == "headToHead"
```

You should already be familiar with the left-hand segment, which selects the `Method` column from the `battleHistory` dataset. By using the equivalency operator (==) and `"headToHead"`, we are telling our function to select only the rows in the `Method` column that contain a value of `headToHead`. In words, this argument can be read as "in the `battleHistory` dataset, select rows from the `Method` column only if they have a value of `headToHead`." Hence, our resulting subset yielded only the 30 rows from our original dataset that contained the head to head combat method.

Pop quiz

1. What does an ellipsis (`...`) mean when encountered inside an R function definition?
   a. The function accepts a single argument.
   b. The function accepts multiple arguments.
   c. The function accepts a specific number of arguments.
   d. The function accepts a variable number of arguments.
2. Interpret the following argument of the `subset(data, ...)` function in words: `battleHistory$Result != "Victory"`
   a. In the `battleHistory` dataset, select rows from column `Result` only if they do not have a value of `Victory`.
   b. In the `battleHistory` dataset, select rows from column `Result` only if they have a value of `Victory`.
   c. In the `battleHistory` dataset, select cells from column `Result` only if they do not have a value of `Victory`.
   d. In the `battleHistory` dataset, select cells from column `Result` only if they have a value of `Victory`.

Have a go hero

Now that you are familiar with extracting information from large datasets, use the `subset(data, ...)` function to create subsets for each of the remaining battle methods—surround, ambush, and fire. Save each of these subsets into new variables, named `subsetSurround`, `subsetAmbush`, and `subsetFire` respectively.

Time for action – deriving summary statistics

A sound way to initiate a deep data analysis is by deriving **summary**, or **descriptive, statistics**. These include simple, although highly informative, calculations such as means, standard deviations, and ranges, amongst others. Summary statistics are excellent for revealing overarching trends and patterns in a dataset. They provide us with a global understanding of our data.

For all calculations, we will store our summary statistics in new variables. For the time being, we will continue to focus on our head to head combat data.

1. Calculate the means, as shown in the following example:

```
> #use mean(data) to calculate the mean of a given dataset
> #what was the mean number of Shu soldiers engaged in past
head to head conflicts?
> meanShuSoldiersHeadToHead <-
mean(subsetHeadToHead$ShuSoldiersEngaged)
> #what was the mean number of Wei soldiers engaged in past
head to head conflicts?
> meanWeiSoldiersHeadToHead <-
mean(subsetHeadToHead$WeiSoldiersEngaged)
> #what was the mean duration (in days) of past head to head
conflicts?
> meanDurationHeadToHead <- mean(subsetHeadToHead$DurationInDays)
```

2. Display each of your mean variables in the R console:

```
> #display the calculated means
> meanShuSoldiersHeadToHead
[1] 31341.67
> meanWeiSoldiersHeadToHead
[1] 33833.33
> meanDurationHeadToHead
[1] 77.93333
```

3. Calculate the standard deviations, and consider the following:

```
> #use sd(data) to calculate the standard deviation of a
given dataset
> #what was the standard deviation of Shu soldiers engaged in past
head to head conflicts?
> sdShuSoldiersHeadToHead <-
sd(subsetHeadToHead$ShuSoldiersEngaged)
> #what was the standard deviation of Wei soldiers engaged in
past head to head conflicts?
```

```
> sdWeiSoldiersHeadToHead <-
sd(subsetHeadToHead$WeiSoldiersEngaged)
> #what was the standard deviation of duration (in days)
in past head to head conflicts?
> sdDurationHeadToHead <- mean(subsetHeadToHead$DurationInDays)
```

4. Display each of your standard deviation variables in the R console:

```
> #display the calculated standard deviations
> sdShuSoldiersHeadToHead
[1] 31320.13
> sdWeiSoldiersHeadToHead
[1] 41192.22
> sdDurationHeadToHead
[1] 77.93333
```

5. Calculate the ranges, as shown in the following:

```
> #use range(data, ...) to calculate the range of a given dataset
> #what was the range of Shu soldiers engaged in past head to
head conflicts?
> rangeShuSoldiersHeadToHead <-
range(subsetHeadToHead$ShuSoldiersEngaged)
> #what was the range of Wei soldiers engaged in past head to
head conflicts?
> rangeWeiSoldiersHeadToHead <-
range(subsetHeadToHead$WeiSoldiersEngaged)
> #what was the range of duration (in days) of past head to
head conflicts?
> rangeDurationHeadToHead <-
range(subsetHeadToHead$DurationInDays)
```

6. Display each of your range variables in the R console:

```
> #display the calculated ranges
> rangeShuSoldiersHeadToHead
[1] 250 100000
> rangeWeiSoldiersHeadToHead
[1] 500 200000
> rangeDurationHeadToHead
[1] 30 120
```

7. Display a general summary of the data:

```
> #use the summary(object) function to generate a summary
of a given object
> #general summary of our head to head combat data
> summaryHeadToHead <- summary(subsetHeadToHead)
```

8. Display your summary variable in the R console. Your values should match the ones pictured in the following screenshot:

```
> #display the head to head subset summary
> summaryHeadToHead
```

```
> summaryHeadToHead
       Method        Rating     SuccessfullyExecuted       Result    ShuSoldiersEngaged WeiSoldiersEngaged DurationInDays
 ambush    : 0   Min.   :10   N: 0                    Defeat :20   Min.   :   250     Min.   :   500     Min.   : 30.00
 fire      : 0   1st Qu.:20   Y:30                    Victory:10   1st Qu.:  5000     1st Qu.: 10000     1st Qu.: 46.00
 headToHead:30   Median :25                                        Median : 25000     Median : 15000     Median : 90.50
 surround  : 0   Mean   :46                                        Mean   : 31342     Mean   : 33833     Mean   : 77.93
                 3rd Qu.:80                                        3rd Qu.: 50000     3rd Qu.: 45000     3rd Qu.:100.00
                 Max.   :95                                        Max.   :100000     Max.   :200000     Max.   :120.00
```

What just happened?

Through summary statistics, we have gained insights on the overall patterns in our data. Let us take a moment to discuss each one individually.

Means

You are already familiar with calculating means from our previous chapter. Here, we looked specifically at the mean soldier engagement and battle durations for past head to head conflicts. Again we see that the Wei forces tend to outnumber the Shu in battle. The average head to head battle has lasted 78 days.

Standard deviations

A **standard deviation** helps to depict the amount of variability present in a collection of data. The `sd(data)` function can be used to calculate the standard deviation of a given dataset. In our soldier engagement data, the Wei army had a higher standard deviation than the Shu army. This indicates that the Wei forces tended to enter battle with a more variable number of soldiers than the Shu forces. Since the Wei army usually outnumbered the Shu in past battles, it is expected that its standard deviation would be larger.

Ranges

The **range** of a dataset is composed of its minimum and maximum values. By using the `range(data)` function in R, we can list the minimum and maximum values of our data in a single command. Similar to the standard deviations, the Wei have a wider range of soldiers engaged than the Shu. This is a predictable outcome considering the Wei forces' larger numbers. The duration of past head to head conflicts ranged from 30 to a 120 days.

Note that individual minimums and maximums can also be calculated using the `min(data)` and `max(data)` functions.

summary(object)

You also employed one of the most useful and versatile functions available to the R language. The `summary(object)` function generates descriptive statistics and other relevant calculations for an object automatically. In our case, the object was a dataset and our descriptive statistics included means, sums, medians, quartiles, minimums, and maximums. The wonderful thing about R's summary function is that it can be used on nearly any object. Depending on the type of object, the summary function will yield output that is relevant to that object. Therefore, it is not only a fast way to get an overall picture of your data, but it can be used in numerous situations. You should use `summary(object)` often, especially when you are beginning to analyze a dataset or want to inspect a newly created object.

Why use summary statistics?

You probably noticed that some of our summary statistic calculations yielded unsurprising and predictable results. This is not, however, reason to discount their value or an argument for abandoning them. In fact, using summary statistics to confirm that our data are *normal* is an essential early step in the data analysis process. In contrast, any value that stands out as peculiar in our summary statistics warrants further inspection. When this occurs, we may have discovered erroneous or outlying data points, or possibly counterintuitive or unforeseen trends and patterns.

For instance, the median duration of head to head battles (91 days) is noticeably higher than the mean duration (78 days). This may indicate that most battles tend to last on the longer side of our 30 to 120 day duration range and that our mean is being skewed downward by a small number of brief battles. By looking back at our head to head subset, we can confirm or deny this observation.

Pop quiz

1. What is the major purpose of the `summary(object)` function in R?
   a. To provide summary statistics relevant to a given variable.
   b. To provide summary statistics relevant to a given dataset.
   c. To provide summary statistics relevant to a given object.
   d. To provide summary statistics relevant to a given subset.
2. Which of the following is not a benefit of summary statistics?
   a. Summary statistics help provide overview information on a dataset.
   b. Summary statistics help answer very detailed questions about a dataset.
   c. Summary statistics help to validate a dataset.
   d. Summary statistics help to expose potential areas of concern and interest within a dataset.

Have a go hero

Now that you are familiar with deriving summary statistics, calculate the means, standard deviations, and ranges for each of the remaining battle methods—surround, ambush, and fire. Also generate a summary of each subset. Follow a similar console structure and naming convention that we used with our head to head combat data. For example, you should create the following variables using your ambush data:

- `meanShuSoldiersAmbush`, `meanWeiSoldiersAmbush`, `meanDurationAmbush`
- `sdShuSoldiersAmbush`, `sdWeiSoldiersAmbush`, `sdDurationAmbush`
- `rangeShuSoldiersAmbush`, `rangeWeiSoldiersAmbush`, `rangeDurationAmbush`
- `summaryAmbush`

Time for action – quantifying categorical variables

Categorical or **nominal data** is information that is classified into nonnumeric levels. Two pertinent columns in our battle history dataset, and subsequently our head to head combat subset, are represented by categorical data. These are the `SuccessfullyExecuted` (categorized as `Y` or `N`) and `Result` (categorized as `Victory` or `Defeat`) columns. A major benefit of categorical data is that it represents information in a very practical and understandable manner. However, categorical data is not well-suited for quantitative data analysis. Fortunately, R is able to recode categorical data in numeric form, thus allowing us to analyze it quantitatively.

Let us proceed through the steps required to recode our `SuccessfullyExecuted` and `Result` columns and save them as numeric variables:

1. Recode the `SuccessfullyExecuted` column using `as.numeric(data)`, as can be seen in the following:

```
> #represent categorical data numerically using as.numeric(data)
> #recode the SuccessfullyExecuted column into N = 1 and Y = 2
> numericExecutionHeadToHead <-
as.numeric(subsetHeadToHead$SuccessfullyExecuted)
```

2. Display the contents of your numeric variable in the R console.

```
> #display the contents of numericSuccessfullyExecutedHeadToHead
> numericExecutionHeadToHead
[1] 2 2 2 2 2 2 2 2 2 2 2 2 2 2 2 2 2 2 2 2 2 2 2 2 2 2 2 2 2 2
```

[Note that if you prefer your categorical variables to begin with a value of zero, as in `N = 0` and `Y = 1`, then you should subtract one from our statement in step 1.]

3. Recode the `SuccessfullyExecuted` column so it begins with a value of zero.

```
> #recode the SuccessfullyExecuted column into N = 0 and Y = 1
> #by default, R recodes variables alphabetically from 1 to n,
so subtract one to offset the coding from 0 to n
> numericExecutionHeadToHead <-
as.numeric(subsetHeadToHead$SuccessfullyExecuted) - 1
```

4. Display the contents of your revised variable in the R console:

```
> #display the contents of numericExecutionHeadToHead
> numericExecutionHeadToHead
[1] 1 1 1 1 1 1 1 1 1 1 1 1 1 1 1 1 1 1 1 1 1 1 1 1 1 1 1 1 1 1
```

5. Recode the `Result` column using `as.numeric(data)`:

```
> #recode the Result column into Defeat = 0 and Victory = 1
> numericResultHeadToHead <- as.numeric(subsetHeadToHead$Result)
- 1
```

6. Display the contents of your numeric variable in the R console:

```
> #display the contents of numericResultHeadToHead
> numericResultHeadToHead
[1] 0 0 0 0 0 0 0 1 1 1 0 0 0 0 0 0 1 1 1 1 0 0 0 0 0 0 0 1 1 1
```

What just happened?

You have represented your categorical columns (`SuccessfullyExecuted` and `Result`) from the head to head combat dataset as numeric variables, thereby preparing them for quantitative analysis. During this process, you encountered the `as.numeric(data)` function and exercised your ability to overwrite variables.

as.numeric(data)

The `as.numeric(data)` function is used to represent nonnumeric data in numeric terms. For example, we used `as.numeric(data)` to convert our `N` and `Y` text values from the `SuccessfullyExecuted` column into the numbers `0` and `1` respectively, using the following:

```
> numericExecutionHeadToHead <-
as.numeric(subsetHeadToHead$SuccessfullyExecuted) - 1
```

Similarly, we used `as.numeric(data)` to code our `Result` column text of `Defeat` and `Victory` into the numbers `0` and `1`:

```
> numericResultHeadToHead <- as.numeric(subsetHeadToHead$Result) - 1
```

Although our data contained only two categories, note that the `as.numeric(data)` function is capable of handling any number of levels. For instance, it would be able to code a variable containing levels for low, medium, and high as 0, 1, and 2.

Overwriting variables

In step 1 of our activity, we originally recoded our `SuccessfullyExecuted` column using values of `N` as 1 and `Y` as 2 and saved the results into a variable called `numericExecutionHeadToHead`, this was done by the following command:

```
> numericExecutionHeadToHead <-
as.numeric(subsetHeadToHead$SuccessfullyExecuted)
```

Then, in step 3, we recoded the column using values of `N` as 0 and `Y` as 1 and then saved the results into a variable with the same name of `numericExecutionHeadToHead`:

```
> numericExecutionHeadToHead <-
as.numeric(subsetHeadToHead$SuccessfullyExecuted) - 1
```

While this was a seamless process that occurred without interruption, it demonstrates an important property of R variables. That is, R variables can be reassigned to new values. When a variable is overwritten in this manner, it assumes a new value and abandons its previous one. So, after step 3, our `numericSuccessfullyExecutedHeadToHead` variable represented `N` and `Y` as 0 and 1 and ceased to depict the values as we had defined them in step 1.

To demonstrate this point, consider variable `A`, which has yet to be assigned a value. Once we execute the line:

```
> A <- 1
```

Variable `A` will take on a value of 1 in the preceding line. If we were then to enter the line:

```
> A <- 2
```

Variable `A` would take on a value of 2 in the preceding line. Its previous contents would be overwritten and therefore forgotten.

Pop quiz

1. What values would represent `N` and `Y` in the `SuccessfullyExecuted` column if it were recoded using the following line?

   ```
   > as.numeric(as.numeric(subsetHeadToHead$SuccessfullyExecuted) + 5
   ```

   a. `N` = 0 and `Y` = 1

   b. `N` = 1 and `Y` = 2

   c. `N` = 5 and `Y` = 6

   d. `N` = 6 and `Y` = 7

2. What would be the value of variable `A` after the following lines were executed in the R console?

   ```
   > A <- 0
   > A <- 1
   > A <- 2
   > A <- 3
   ```

   a. 3

   b. 2

   c. 1

   d. 0

Have a go hero

Now that you have quantified your first categorical variables, proceed to recode the `SuccessfullyExecuted` and `Result` columns for each of the remaining battle methods—surround, ambush, and fire. Follow a similar console structure and naming convention that we used with our head to head combat data. For example, you should create the following variables with your ambush data:

- `numericExecutionAmbush`
- `numericResultAmbush`

Time for action – correlating variables

Correlations tell us how well two variables relate to each other. As with summary statistics, calculating the correlations between variables in our dataset is a fast and easy way to acquire an initial understanding of our data.

Let us use correlations to investigate a few of the relationships in our head to head battle data:

1. Calculate the correlation between `Rating` and `Result`. Be sure to use the numeric version of the `Result` column in your calculation:

```
> #use cor(x,y) to calculate the correlation between two variables
> #remember only to use numeric values when calculating
correlations
> #How is the performance rating of the Shu army related to the
outcome of a head to head battle?
> corRatingResultHeadToHead <- cor(subsetHeadToHead$Rating,
numericResultHeadToHead)
```

2. Display the value of your correlation in the R console:

```
> #display the value of the correlation
> corRatingResultHeadToHead
[1] 0.9495232
```

3. Calculate the correlation between `ShuSoldiersEngaged` and `WeiSoldiersEngaged`:

```
> #How is the number of Shu soldiers engaged in a head to head
battle correlated with the number of Wei soldiers engaged?
> corShuWeiSoldiersHeadToHead <-
cor(subsetHeadToHead$ShuSoldiersEngaged,
subsetHeadToHead$WeiSoldiersEngaged)
```

4. Display the value of your correlation in the R console:

```
> #display the value of the correlation
> corShuWeiSoldiersHeadToHead
[1] 0.7653596
```

5. Calculate the correlations between (almost) all of the variables in the dataset:

```
> use cor(data) to calculate the correlation between all
numeric variables in a dataset
> #How are all of our numeric battle data correlated with
one another?
> corHeadToHead <- cor(subsetHeadToHead)
```

6. Display the values of your correlations in the R console, by using the following:

```
> #display the correlations
> corHeadToHead
```

> corHeadToHead

	Method	Rating	SuccessfullyExecuted	Result	ShuSoldiersEngaged	WeiSoldiersEngaged	DurationInDays
Method	1	NA	NA	NA	NA	NA	NA
Rating	NA	1.00000000	NA	NA	0.4222706	-0.04668125	-0.8785341
SuccessfullyExecuted	NA	NA	1	NA	NA	NA	NA
Result	NA	NA	NA	1	NA	NA	NA
ShuSoldiersEngaged	NA	0.42227061	NA	NA	1.0000000	0.76535963	-0.2356156
WeiSoldiersEngaged	NA	-0.04668125	NA	NA	0.7653596	1.00000000	0.1378893
DurationInDays	NA	-0.87853412	NA	NA	-0.2356156	0.13788932	1.0000000

What just happened?

We calculated just a few correlations to get an idea of how they can be derived in R. This entailed using the `cor()` function in two different ways.

Interpreting correlations

Correlations range in value from negative one (-1) to positive one (1). A value of negative one means that two variables are perfectly negatively correlated. That is, a high value in one is associated with a low value in the other, and vice versa. On the other hand, a correlation of positive one indicates that two variables are perfectly positively correlated. As such, high values in one are associated with high values in the other, and vice versa. Further, a correlation of zero indicates that two variables are perfectly uncorrelated. This means that their values do not associate with one another. Of course, these extreme correlational values are rare. Most correlations will fall somewhere between negative one and zero or zero and positive one.

Here are a few examples that demonstrate how to interpret correlations:

- A correlation of 0.12 between `A` and `B` suggests a relatively weak positive relationship exists between the variables. If `A` were to decrease by a certain amount, we would only expect a small decrease in `B`.
- A correlation of -0.87 between `A` and `B` suggests a relatively strong negative relationship exists between the variables. If `A` were to increase, we would expect a `B` to decrease by proportionally similar amount.
- A correlation of 0.00001 between `A` and `B` suggests that the variables are uncorrelated. Therefore, movements in `A` would not be expected to associate with movements in `B`.

An important final note on correlation is that it should never be interpreted as **causation**. Correlation merely tells us that our variables tend to move with each other in a certain way. Yet, we cannot determine which, if either, of the correlated variables causes the change in the other. Therefore, correlations inform us about *what* is occurring between our variables, but cannot tell us *why* it is happening.

cor(x, y)

The `cor(x,y)` function is used to calculate the correlation between two variables, `x` and `y`. For instance, to calculate the correlation between variable `A` and variable `B`, we would use the following code:

```
> cor(A, B)
```

We looked directly at two correlations. First, we found the correlation between the performance rating of the Shu army and the outcome of head to head battles to be 0.95. This correlation suggests that victory or defeat in a given head to head battle had a large impact on Zhuge Liang's rating of the army's performance in that conflict.

Next, we calculated the correlation between the number of Shu and Wei soldiers engaged in head to head battles. Here, we found a relatively strong positive correlation of 0.77. This suggests that the number of soldiers that one army engages in combat is highly related to the size of the opposing army. This is logical, because we would expect an army's size in a given battle to be closely related to (but not necessarily determined by or equal to) the size of the opposing army.

cor(data)

The same correlation function can be used in a different way. Instead of providing `x` and `y` variables to calculate a single correlation via `cor(x, y)`, we can calculate all of the possible correlations in a dataset using `cor(data)`. For example, to find the correlations for all of the numeric variables in dataset `A`, we would use the following code:

```
> cor(A)
```

This use of the `cor()` function yields a **correlation table**, similar to the one that we generated for our head to head dataset.

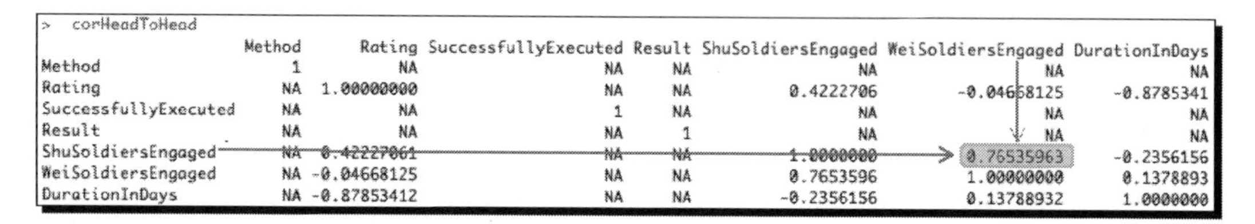

> corHeadToHead

	Method	Rating	SuccessfullyExecuted	Result	ShuSoldiersEngaged	WeiSoldiersEngaged	DurationInDays
Method	1	NA	NA	NA	NA	NA	NA
Rating	NA	1.00000000	NA	NA	0.4222706	-0.04668125	-0.8785341
SuccessfullyExecuted	NA	NA	1	NA	NA	NA	NA
Result	NA	NA	NA	1	NA	NA	NA
ShuSoldiersEngaged	NA	0.42227061	NA	NA	1.0000000	0.76535963	-0.2356156
WeiSoldiersEngaged	NA	-0.04668125	NA	NA	0.7653596	1.00000000	0.1378893
DurationInDays	NA	-0.87853412	NA	NA	-0.2356156	0.13788932	1.0000000

To read a value from this table, match a row name on the left-hand side with a column name across the top. At the intersection, you will find the correlation between the two variables. For instance, if you traced from `ShuSoldiersEngaged` on the left to `WeiSoldiersEngaged` on the top, you would encounter the correlation of 0.77 that we had previously calculated using `cor(x,y)`.

NA values

A critical limitation of the `cor(data)` technique is that only numeric variables in a dataset can be correlated. You probably noticed that several `NA` values were reported in the correlation table of our head to head dataset. These occur because our `SuccessfullyExecuted` and `Result` columns consisted of nonnumeric data. Therefore they could not be correlated and R returned `NA` values. To correlate nonnumeric values, as we did with `Result` in step 1, they must first be recoded as numeric.

[See the *Quantifying categorical variables* section of this chapter for a demonstration of how to recode nonnumeric data in numeric form.]

You may run into `NA` values in other aspects of your R work. When these occurs, it is a good idea to check your data to make sure that they are in the proper format for the function or calculation that you wish to employ.

Pop quiz

1. What is the key difference between `cor(x,y)` and `cor(data)`?
   a. The `cor(x, y)` variation calculates all of the correlations in a dataset, whereas `cor(data)` calculates a single correlation between two variables.
   b. The `cor(x, y)` variation calculates a single correlation between two variables, whereas `cor(data)` calculates all of the correlations in a dataset.
   c. The `cor(x, y)` variation calculates all of the correlations between two datasets, whereas `cor(data)` calculates all of the correlations in a given dataset.
   d. The `cor(x, y)` variation calculates all correlations between two variables, whereas `cor(data)` calculates all correlations for a given variable.
2. Interpret a correlation of -0.25 between the variables `A` and `B`.
   a. `A` and `B` are negatively correlated. For every one unit increase in `A`, `B` will decrease by 0.25 units.
   b. `A` and `B` are negatively correlated. For every one unit decrease in `A`, `B` will decrease by 0.25 units
   c. `A` and `B` are negatively correlated. We would expect an increase in `A` to be accompanied by a proportionally small increase in `B`.
   d. `A` and `B` are negatively correlated. We would expect an increase in `A` to be accompanied by a proportionally small decrease in `B`.

Have a go hero

You may have noticed that all of the points in our head to head combat dataset have a value of `Y` for `SuccessfullyExecuted`, which prevents us from correlating it with other variables. This is because the Shu forces can engage in head to head combat at will and without some variation in the values for execution, a correlation is incalculable.

In contrast, our surround, ambush, and fire attack methods greatly depend on successful execution. Try correlating the `Rating` column with the `SuccessfullyExecuted` column in each of these battle methods. Then, interpret your findings.

Afterwards, use `cor(data)` to visualize all of the correlations in your datasets. Interpret these correlations and take note of any that stand out as expected or unexpected. By investigating correlations, you are becoming ever more aware of you data.

Regression

We can use **regression analysis** to inform our predictions. Regression analysis is a data modeling technique that helps us understand how different variables change with one another. A **regression model** must incorporate at least one **dependent** (or outcome) and **independent** (or predictor) variable, although several of both can be included. We can use regression models to predict outcomes based on the data that is available to us. As the Shu strategist, you will be predicting your army's performance in battle across different courses of action, based on what you know about past conflicts.

Note that the regression models we create will predict the Shu army's future performance rating based on several conditions of battle. Recall that the past battle ratings were recorded by Zhuge Liang, who was an expert in assessing the army's combat performance. Therefore, Zhuge Liang's ratings can be considered a valid measure for predicting future performance, since they are equivalent to the actual performance of the Shu forces.

While it is technically possible to directly predict the result of battle (victory or defeat) from our dataset, this would require the use of advanced regression techniques that are beyond the scope of this book. Thus, we will focus on predicting outcomes based on performance ratings via the most common regression methods.

Time for action – modelling with simple linear regression

Simple linear regression is the most basic form of regression analysis. It uses a single independent variable to predict the outcome of a single dependent variable.

To begin experimenting with regression analysis in R, let us create a simple linear model from our head to head combat data:

1. Use the `lm(formula, data)` function to create a linear regression model where `Rating` is the dependent variable and `ShuSoldiersEngaged` is the independent variable. This is done as follows:

```
> #create a linear regression model using the lm(formula, data)
> #predict the rating of a head to head battle using the number
of Shu soldiers engaged
> lmHeadToHeadRating_ShuSoldiers <- lm(subsetHeadToHead$Rating ~
subsetHeadToHead$ShuSoldiersEngaged, subsetHeadToHead)
```

2. Display the contents of your linear model variable in the R console:

```
> #display the contents of the model
> lmHeadToHeadRating_ShuSoldiers
```

```
> lmHeadToHeadRating_ShuSoldiers

Call:
lm(formula = subsetHeadToHead$Rating ~ subsetHeadToHead$ShuSoldiersEngaged,     data = subsetHeadToHead)

Coefficients:
                        (Intercept)  subsetHeadToHead$ShuSoldiersEngaged
                          3.146e+01                            4.374e-04
```

3. Create a summary of the model, as follows:

```
> #create the model summary
> lmHeadToHeadRating_ShuSoldiers_Summary <-
summary(lmHeadToHeadRating_ShuSoldiers)
```

4. Display the contents of your linear model summary in the R console:

```
> #display the model summary
> lmHeadToHeadRating_ShuSoldiers_Summary
```

```
> lmHeadToHeadRating_ShuSoldiers_Summary

Call:
lm(formula = subsetHeadToHead$Rating ~ subsetHeadToHead$ShuSoldiersEngaged,
    data = subsetHeadToHead)

Residuals:
    Min      1Q  Median      3Q     Max
-45.199 -23.371  -9.112  24.253  51.981

Coefficients:
                                     Estimate Std. Error t value Pr(>|t|)
(Intercept)                         3.146e+01  7.797e+00   4.035 0.000383 ***
subsetHeadToHead$ShuSoldiersEngaged 4.374e-04  1.774e-04   2.465 0.020098 *
---
Signif. codes:  0 '***' 0.001 '**' 0.01 '*' 0.05 '.' 0.1 ' ' 1

Residual standard error: 29.93 on 28 degrees of freedom
Multiple R-squared: 0.1783,	Adjusted R-squared: 0.149
F-statistic: 6.076 on 1 and 28 DF,  p-value: 0.0201
```

What just happened?

Your first linear regression model yielded quite a bit of information. Let us look at how to use the `lm(formula, data)` function as well as how to interpret the information that it provides to us.

lm(formula, data)

The `lm(formula, data)` function is used to create a linear regression model. The `formula` argument takes on the following structure:

```
dVar ~ iVar1 + iVar2 + ... + iVarn
```

Here, `dVar` is the dependent variable and `iVar1` through `iVarn` are independent variables. While our initial model used a single independent variable, the linear model function is capable of accepting as many as we need. The `data` argument contains the dataset from which our variables are taken. Hence, the basic composition of the `lm(formula, data)` function resembles the following:

```
lm(dVar ~ iVar1 + iVar2 + ... + iVarn, data)
```

In our simple linear regression model, `Rating` acted as the dependent variable and `ShuSoldiersEngaged` took on the role of the independent variable, as shown:

```
> lmHeadToHeadRating_ShuSoldiers <- lm(subsetHeadToHead$Rating ~
subsetHeadToHead$ShuSoldiersEngaged, subsetHeadToHead)
```

Linear model output

Together, we formed a linear model that regressed the Shu army's head to head combat performance rating (the dependent or predicted variable) on the number of Shu soldiers engaged in battle (the independent or predictor variable). When we called our linear model variable, we received the following output from the R console:

```
> lmHeadToHeadRating_ShuSoldiers

Call:
lm(formula = subsetHeadToHead$Rating ~ subsetHeadToHead$ShuSoldiersEngaged,     data = subsetHeadToHead)

Coefficients:
                        (Intercept)  subsetHeadToHead$ShuSoldiersEngaged
                          3.146e+01                            4.374e-04
```

This output consists of two sections. In **Call:**, we see a reiteration of the console line that R used to create the model. In **Coefficients:**, we see both an **intercept** and a **coefficient** for the number of Shu soldiers engaged. The latter two items help us to create a **regression equation**. Typically, a regression equation takes on the following form:

```
Y = b0 + b1X1 + b2X2 + ... + bnXn
```

In this equation, `Y` is the dependent variable, `b0` is the intercept, and `b1X1` through `bnXn` are independent variables. Thus, the equation for our model is as follows:

```
Rating = 31 + 0.00044 * number of Shu soldiers
```

Linear model summary

After displaying the model output, you also created a more detailed summary using the `summary(object)` function

```
> lmHeadToHeadRating_ShuSoldiers_Summary <-
summary(lmHeadToHeadRating_ShuSoldiers)
```

[For a discussion of the `summary(object)` function, revisit the *Deriving summary statistics* section of this chapter.]

```
> lmHeadToHeadRating_ShuSoldiers_Summary

Call:
lm(formula = subsetHeadToHead$Rating ~ subsetHeadToHead$ShuSoldiersEngaged,
    data = subsetHeadToHead)

Residuals:
    Min      1Q  Median      3Q     Max
-45.199 -23.371  -9.112  24.253  51.981

Coefficients:
                                     Estimate Std. Error t value Pr(>|t|)
(Intercept)                         3.146e+01  7.797e+00   4.035 0.000383 ***
subsetHeadToHead$ShuSoldiersEngaged 4.374e-04  1.774e-04   2.465 0.020098 *
---
Signif. codes:  0 '***' 0.001 '**' 0.01 '*' 0.05 '.' 0.1 ' ' 1

Residual standard error: 29.93 on 28 degrees of freedom
Multiple R-squared: 0.1783, Adjusted R-squared: 0.149
F-statistic: 6.076 on 1 and 28 DF,  p-value: 0.0201
```

Again, you have witnessed the value and versatility of the `summary(object)` function, as it adapted itself to generate output relevant to our regression model. In the output, you can see the same intercept and independent variable coefficients (**Estimate** column) that we derived from the default model output. However, you are also exposed to a wealth of additional information about the model. In fact, nearly everything you would need to know for a data analysis is included. For our interpretations, we will focus on the **Coefficients:**, **Multiple R-squared**, and **p-value/Pr(>|t|)** portions of the output.

In case you need to be refreshed on the meaning of **R-squared** and **p-values**, we will briefly review them here:

- R-squared (**Multiple R-squared** in the summary output) tells us how well our linear model fits our data, and thus, how much predictive power our model has. Technically, it is the percentage of variance in the dependent variable that is accounted for by a regression model. For example, the R-squared of our linear model tells us how much of the variance in the performance rating of a head to head conflict can be accounted for by the number of Shu soldiers engaged in that battle.
- A p-value (**Pr(>|t|)** and **p-value** in the summary output) is an indicator of **statistical significance**. In common practice, a cutoff 0.05 is used to determine statistical significance. Both individual coefficients and the overall linear model have p-values. In general, it is better to have significant coefficients and models, because statistical significance indicates that our results are more likely to be genuine and unlikely to have occurred by random chance. Yet, statistical significance is not the be all and end all of data analysis. Since data do not think nor act, one must always remember to consider the practical implications of statistical findings. We will also remain diligent in assessing the **practical significance** of our work throughout this book.

Interpreting a linear regression model

Sound interpretation is essential to understanding the practical ramifications of our data analyses. Recall that our linear regression analysis yielded the following equation:

```
Y = 31 + 0.00044 * X1
```

Or in words:

```
Rating = 31 + 0.00044 * number of Shu soldiers
```

Look back at the `Rating` column of our original battle history dataset. `Rating` can take on a value between 0 and 100. Since we are interested in predicting the Shu army's performance, the closer our equation comes to 100, the more confident we will be that our battle plans will lead to victory. Conversely, the lower our predicted performance, the less confident we can be that our strategy is going to lead to beneficial outcomes.

In fact, it is clear from our data that Zhuge Liang rated the army's performance at or above 80 in victorious battles, whereas he rated the army lower in conflicts that resulted in defeat. Therefore, 80 is a good rating threshold to keep in mind when predicting future battle performance. In general, we want to devise strategies that will predict a performance of 80 or higher.

A model's intercept is interpreted as the value of the dependent variable when all independent variables are equal to zero. The intercept of a linear regression model often does not have an intuitive meaning. In our case, the intercept of 31 suggests that our performance will somehow be greater than zero even if we do not send soldiers into battle. Nevertheless, the intercept impacts our overall model and is important for making predictions.

Our coefficient for the number of Shu soldiers engaged is 0.00044. As you can imagine, it would take quite a large force to predict a sufficient performance rating for victory using our model. This notion is demonstrated by the following calculation, which solves for the number of soldiers necessary to predict a rating of 80:

```
80 = 31 + 0.00044 * X1
49 = 0.00044 * X1
X1 = 111,364 soldiers needed to predict victory!
```

This suggests that over half of the entire Shu army of 200,000 would need to participate in a single battle just to reach our minimum rating threshold. Yet, recall that our current model only deals with head to head combat performance and only uses the number of Shu soldiers engaged to predict it.

While both our coefficient and overall model are statistically significant with p-values of 0.02, there is much that is left unexplained. This is evident when considering our R-squared value of 0.18. This value means that only 18% of the variance in performance rating can be explained by our model. In a practical sense, this can be interpreted as saying that only 18% of the rating of a head to head battle can be accounted for by the number of Shu soldiers engaged.

All in all, our interpretations indicate that the current model is not effective enough at predicting the Shu army's performance. Clearly, there are many other factors that account for performance besides the number of soldiers that we send into battle. Thankfully, we have a dataset that contains rich battle history information and the ability to form more complex multiple linear regression models. Thus, the analysis of our battle data has just begun.

Pop quiz

1. Which of the following represents proper syntax for use in the formula argument of the `lm(formula, data)` function?

   a. `Y ~ X1 - X2`

   b. `Y ~ X1 + X2`

   c. `X ~ Y1 + Y2`

   d. `X ~ Y1 - Y2`

2. In the following linear regression equation, identify the dependent variable, independent variable, intercept, and coefficient: `Y = 0.5 + 3 * X`
   a. `Y` is the dependent variable, `X` is the independent variable, `0.5` is the intercept, and `3` is the coefficient.
   b. `X` is the dependent variable, `Y` is the independent variable, `0.5` is the intercept, and `3` is the coefficient.
   c. `Y` is the dependent variable, `X` is the independent variable, `3` is the intercept, and `0.5` is the coefficient.
   d. `Y` is the dependent variable, `3` is the independent variable, `0.5` is the intercept, and `X` is the coefficient.
3. Interpret the following linear regression equation: `Y = 5 - 10 * X`
   a. The predicted value of `Y` is equal to `5` plus `10` times `X`.
   b. The value of `Y` is equal to `5` plus `10` times `X`.
   c. The predicted value of `Y` is equal to `5` minus `10` times `X`.
   d. The value of `Y` is equal to `5` minus `10` times `X`.

Time for action – modelling with multiple linear regression

Multiple linear regression is one step removed from simple linear regression. It adheres to the sample principles, but makes use of additional independent variables to predict the outcome of a dependent variable.

Let us build upon our previous head to head combat model using multiple regression. This time, we will include both the number of Shu and Wei soldiers engaged as predictors of battle performance:

1. Create a multiple regression model that predicts `Rating` using both the number of Shu and Wei soldiers engaged:

```
> #create a multiple linear regression model using the
lm(formula, data) function
> #predict the rating of a head to head battle using the number
of Shu and Wei soldiers engaged
> lmHeadToHeadRating_ShuWeiSoldiers <- lm(subsetHeadToHead$Rating
~ subsetHeadToHead$ShuSoldiersEngaged +
subsetHeadToHead$WeiSoldiersEngaged, subsetHeadToHead)
```

2. Create a summary of the model:

```
> #model summary
> lmHeadToHeadRating_ShuWeiSoldiers_Summary <-
summary(lmHeadToHeadRating_ShuWeiSoldiers)
```

3. Display your linear model summary in the R console:

```
> #display the summary
> lmHeadToHeadRating_ShuWeiSoldiers_Summary
```

```
> lmHeadToHeadRating_ShuWeiSoldiers_Summary

Call:
lm(formula = subsetHeadToHead$Rating ~ subsetHeadToHead$ShuSoldiersEngaged +
    subsetHeadToHead$WeiSoldiersEngaged, data = subsetHeadToHead)

Residuals:
    Min      1Q  Median      3Q     Max
-29.765 -17.896  -6.758  14.918  46.789

Coefficients:
                                       Estimate Std. Error t value Pr(>|t|)
(Intercept)                          33.0642804  6.1518218   5.375 1.11e-05 ***
subsetHeadToHead$ShuSoldiersEngaged   0.0011453  0.0002171   5.275 1.46e-05 ***
subsetHeadToHead$WeiSoldiersEngaged  -0.0007033  0.0001651  -4.260 0.000222 ***
---
Signif. codes:  0 '***' 0.001 '**' 0.01 '*' 0.05 '.' 0.1 ' ' 1

Residual standard error: 23.57 on 27 degrees of freedom
Multiple R-squared: 0.5086, Adjusted R-squared: 0.4722
F-statistic: 13.97 on 2 and 27 DF,  p-value: 6.833e-05
```

What just happened?

We used multiple linear regression to create a second model for predicting the performance rating of the Shu army in a head to head conflict. This model incorporated both the number of Shu and number of Wei soldiers engaged in combat as predictors. We can interpret a multiple linear regression model in a similar manner to a simple linear regression model. We can also compare our new model to the one that we previously created.

Interpreting the summary output

Review the summary output for our multiple regression model. The summary should be similar to the following screenshot:

```
> lmHeadToHeadRating_ShuWeiSoldiers_Summary

Call:
lm(formula = subsetHeadToHead$Rating ~ subsetHeadToHead$ShuSoldiersEngaged +
    subsetHeadToHead$WeiSoldiersEngaged, data = subsetHeadToHead)

Residuals:
    Min      1Q  Median      3Q     Max
-29.765 -17.896  -6.758  14.918  46.789

Coefficients:
                                       Estimate Std. Error t value Pr(>|t|)
(Intercept)                          33.0642804  6.1518218   5.375 1.11e-05 ***
subsetHeadToHead$ShuSoldiersEngaged   0.0011453  0.0002171   5.275 1.46e-05 ***
subsetHeadToHead$WeiSoldiersEngaged  -0.0007033  0.0001651  -4.260 0.000222 ***
---
Signif. codes:  0 '***' 0.001 '**' 0.01 '*' 0.05 '.' 0.1 ' ' 1

Residual standard error: 23.57 on 27 degrees of freedom
Multiple R-squared: 0.5086, Adjusted R-squared: 0.4722
F-statistic: 13.97 on 2 and 27 DF,  p-value: 6.833e-05
```

From the **Estimate** column, we can derive our regression equation:

```
Rating = 33 + 0.0011 * ShuSoldiersEngaged - 0.00007 *
WeiSoldiersEngaged
```

Again, both our overall model ($p < .001$) and our independent variable coefficients ($p < .001$) are statistically significant. Moreover, the R-squared increased compared to our previous model to explain 51% of the variance in the Shu army's performance rating.

Let us use our multiple regression model to predict the performance of a 25,000 soldier Shu army against a 25,000 soldier Wei army, as follows:

```
Rating = 33 + 0.0011 * 25000 - 0.00007 * 25000
Rating = 33 + 27.5 - 1.75
Rating = 58.75
```

Recall that our `Rating` variable ranges from 0 to 100 and that our past victories have achieved ratings of 80 or higher. Our predicted rating of 59 suggests that the Shu army would likely not be victorious in this hypothetical conflict. However, also recall that our model only contains 51% of the ingredients that account for changes in head to head battle performance. Furthermore, our initial inference at the beginning of this chapter revealed that the Wei forces tend to enter a given battle with many more soldiers than the Shu. For these reasons, our model, as well as our hypothetical example, may not have sufficient practical relevance.

Explaining model differences

The increase in R-squared from our simple regression model to our multiple regression model can be attributed to the fact that our new model included more information that is relevant to predicting head to head battle performance. Our multiple regression model factors in the size of both armies when determining the Shu army's rating. Since the ability of the Shu army to perform well is dependent to some extent on the opposing forces, including both armies yields a much stronger basis for prediction than the single army approach that our original model took.

The key to developing useful predictive regression models is to include only the most relevant data. While 51% is a large improvement in predictive power over our preceding model, it still may not be enough to make us confident in making critical strategy decisions for the Shu army. Certainly, we are encouraged to explore the full range of our data before settling on a particular model.

Pop quiz

1. Which of the following is most likely to increase the statistical significance of a multiple regression model?
    a. Including more independent variables.
    b. Including fewer independent variables.
    c. Including more relevant and fewer irrelevant independent variables.
    d. Including more irrelevant and fewer relevant independent variables.
2. Which of the following is most likely to increase the practical significance of a multiple regression model?
    a. Including more independent variables.
    b. Including fewer independent variables.
    c. Including more relevant and fewer irrelevant independent variables.
    d. Including more irrelevant and fewer relevant independent variables.

Have a go hero

Create a new simple linear regression model that uses `DurationInDays` alone to predict the Shu army's performance in a head to head conflict. Then create two new multiple linear regression models that expand upon the previous model by incorporating `ShuSoldiersEngaged` and `WeiSoldiersEngaged` respectively. Generate and interpret the model summaries. Once complete, you should have three new regression models:

- `lmHeadToHeadRating_Duration`
- `lmHeadToHeadRating_DurationShuSoldiers`
- `lmHeadToHeadRating_DurationSoldiers`

Also there should be three accompanying summaries saved in your R workspace. What do these models tell you about the importance of the duration of battle in predicting the outcome of head to head conflicts?

Time for action – modelling interactions

One other way that we can explore the relationships in our data is by looking at **interaction effects**. An interaction spawns from an interplay between variables whereby the interaction effect is different from either of the variables alone. **Interaction variables** can be created in R, although a specific procedure must be followed to use them properly.

Let us look at how an interaction variable can be created and incorporated into a regression model in R. We will accomplish this by including the interaction between Shu and Wei soldiers engaged as a variable in our multiple regression model:

1. Center the two variables that you plan to interact:

```
> #before creating an interaction variable, the component
variables must first be centered
> #center a variable by subtracting its mean from each of its
values
> #center the number of Shu soldiers engaged
> centeredShuSoldiersHeadToHead <-
subsetHeadToHead$ShuSoldiersEngaged -
mean(subsetHeadToHead$ShuSoldiersEngaged)
> #center the number of Wei soldiers engaged
> centeredWeiSoldiersHeadToHead <-
subsetHeadToHead$WeiSoldiersEngaged -
mean(subsetHeadToHead$WeiSoldiersEngaged)
```

2. Multiply the two centered variables to create the interaction variable:

```
> #create an interaction variable by multiplying two or more
centered variables
> interactionSoldiersHeadToHead <-
centeredShuSoldiersHeadToHead * centeredWeiSoldiersHeadToHead
```

3. Create an interaction model that predicts `Rating` using the duration, Shu soldiers engaged, Wei soldiers engaged, and the interaction between the number of Shu and Wei soldiers engaged:

```
> #predict the rating of a battle using the duration, number of
Shu and Wei soldiers engaged, and the interaction between the
number of Shu and Wei soldiers engaged
> lmHeadToHeadRating_DurationSoldiersShuWeiInteraction <-
lm(subsetHeadToHead$Rating ~  subsetHeadToHead$DurationInDays +
subsetHeadToHead$ShuSoldiersEngaged +
subsetHeadToHead$WeiSoldiersEngaged +
interactionSoldiersHeadToHead, subsetHeadToHead)
```

4. Create a summary of the model:

```
> #model summary
> lmHeadToHeadRating_DurationSoldiersShuWeiInteraction_Summary
<-
summary(lmHeadToHeadRating_DurationSoldiersShuWeiInteraction)
```

5. Display your interaction model summary in the R console:

```
> #display the summary
> lmHeadToHeadRating_DurationSoldiersShuWeiInteraction_Summary
```

```
> lmHeadToHeadRating_DurationSoldiersShuWeiInteraction_Summary

Call:
lm(formula = subsetHeadToHead$Rating ~ subsetHeadToHead$DurationInDays +
    subsetHeadToHead$ShuSoldiersEngaged + subsetHeadToHead$WeiSoldiersEngaged +
    interactionSoldiersHeadToHead, data = subsetHeadToHead)

Residuals:
     Min       1Q   Median       3Q      Max
-25.3749  -8.9428   0.6675   9.3589  22.2165

Coefficients:
                                      Estimate Std. Error t value Pr(>|t|)
(Intercept)                          9.956e+01  8.871e+00  11.222 2.98e-11 ***
subsetHeadToHead$DurationInDays     -7.977e-01  9.890e-02  -8.066 2.02e-08 ***
subsetHeadToHead$ShuSoldiersEngaged  4.661e-04  1.553e-04   3.000  0.00603 **
subsetHeadToHead$WeiSoldiersEngaged -1.338e-04  1.635e-04  -0.819  0.42078
interactionSoldiersHeadToHead       -2.413e-09  2.143e-09  -1.126  0.27090
---
Signif. codes:  0 '***' 0.001 '**' 0.01 '*' 0.05 '.' 0.1 ' ' 1

Residual standard error: 12.83 on 25 degrees of freedom
Multiple R-squared: 0.8653, Adjusted R-squared: 0.8437
F-statistic: 40.14 on 4 and 25 DF,  p-value: 1.549e-10
```

What just happened?

You have completed the process of creating and implementing an interaction variable. The resulting interaction model expanded upon our multiple regression model by factoring in the the interplay between the number of Shu and Wei soldiers on the performance rating of the Shu army. Let us review the two-step interaction variable creation process and discuss how such variables can be interpreted:

1. Center the input variables:

   The initial step in creating an interaction variable is to **center** the input variables that you wish to interact. This is accomplished by subtracting the mean of all of the values from each data point. For example, in:

   ```
   centeredA <- A - mean(A)
   ```

   The centered version of variable `A` is created by subtracting the mean of `A` from each value of `A`.

   Centering is necessary because it mitigates the threat of **multicollinearity**, which occurs when two or more independent variables are highly correlated with one another. For instance, our interaction variable was composed of the number of Shu and Wei soldiers engaged in head to head combat. At the same time, our regression model used these variables as separate predictors. Naturally, multicollinearity is a threat in this situation, because our interaction variable is composed of the same data as our other predictors. Thankfully, the centering process is effective in mitigating most of the ill-effects that can be attributed to multicollinearity.

2. Multiply the input variables:

   The second step in creating an interaction variable is to multiply the centered versions of the input variables, like so:

   ```
   interactionAB <- centeredA * centeredB
   ```

   Afterwards, your interaction variable can be used in the same manner as any other variable within a regression model.

Interpreting interaction variables

The statistical significance of the interaction coefficient is an indication of whether or not an interaction is present in the data. When present, an interaction suggests that the relationship between the dependent variable and a predictor varies as the value of the interacting predictor (Wei soldiers) changes. This phenomenon is sometimes referred to as a **moderation effect**, because it describes how one predictor moderates, or affects the strength or direction of, the relationship between another predictor and the dependent variable. When an interaction is absent, the relationship between the dependent variable and a given predictor is not believed to alter as the value of the interacting predictor changes.

The interaction term in our latest model was not statistically significant and did not increase the predictive power of the model. This is logical in our situation. If there were an interaction, then we would expect the number of soldiers that one side engaged to differ across the range of soldiers that the other side deployed. For example, if the Shu engaged 1000 soldiers in battle, then the Wei might deploy 10000 (ten times), but if the Shu engaged 10000, the Wei might deploy 500000 (fifty times). In contrast, without the interaction, we would not expect the number of soldiers engaged by one side to vary across the range of soldiers that the other side deployed. Furthermore, the number of soldiers deployed may be better explained by situational attributes, such as the number of soldiers that happen to be available at a given place or time when a battle occurs. The latter explanations have more practical meaning than the interaction interpretation and help to verify the absence of an interaction effect.

Pop quiz

1. How is a variable centered in R?
    a. By adding its mean to each of its values.
    b. By subtracting its mean from each of its values.
    c. By multiplying its mean by each of its values.
    d. By dividing its mean by each of its values.
2. How is an interaction variable created in R?
    a. By adding the two variables that are believed to interact.
    b. By multiplying the two variables that are believed to interact.
    c. By adding the centered versions of the two variables that are believed to interact.
    d. By multiplying the centered versions of the two variables that are believed to interact.
3. Which of the following would be a viable interpretation of a statistically significant interaction between the variables `A` and `B`?
    a. The relationship between `B` and the dependent variable fluctuates based on the value of `A`.
    b. The relationship between `A` and `B` fluctuates based on the value of the dependent variable.
    c. The value of the dependent variable fluctuates based on the relationship between `A` and `B`.
    d. The value of the dependent variable fluctuates based on the values of `A` and `B`.

Have a go hero

Consider the data in one of your remaining battle method subsets (surround, ambush, or fire). Use the techniques that we have employed in this chapter to create a multiple regression model that incorporates an interaction variable. Then interpret the model. Be sure to address the meaning and significance of the interaction that you explored.

Time for action – comparing and choosing models

At the moment, we have several models that attempt to predict the performance rating of the Shu army in head to head battles based on the duration and number of soldiers engaged in that battle. Yet, we do not have answers regarding which model is best and the relative contribution that each model makes above and beyond the preceding models.

We can use the process of **hierarchical linear regression (HLR)** to compare our models. Let us look at how HLR can be used to compare the three models that we have made thus far:

1. Display a summary of each model:

```
> #use HLR to compare different models
> #first consider the models individually
> #simple regression model using duration to predict battle
rating
> lmHeadToHeadRating_Duration_Summary
```

This should produce a result as shown in the following screenshot:

```
> lmHeadToHeadRating_Duration_Summary

Call:
lm(formula = subsetHeadToHead$Rating ~ subsetHeadToHead$DurationInDays,
    data = subsetHeadToHead)

Residuals:
    Min      1Q  Median      3Q     Max
-36.328  -9.003   2.853   8.252  25.894

Coefficients:
                                Estimate Std. Error t value Pr(>|t|)
(Intercept)                    119.54682    8.16731  14.637 1.20e-14 ***
subsetHeadToHead$DurationInDays  -0.95441    0.09807  -9.732 1.75e-10 ***
---
Signif. codes:  0 '***' 0.001 '**' 0.01 '*' 0.05 '.' 0.1 ' ' 1

Residual standard error: 15.77 on 28 degrees of freedom
Multiple R-squared: 0.7718, Adjusted R-squared: 0.7637
F-statistic: 94.71 on 1 and 28 DF,  p-value: 1.747e-10
```

```
> #multiple regression model using duration, Shu soldiers, and Wei
soldiers to predict battle rating
> lmHeadToHeadRating_DurationSoldiers_Summary
```

This should give a summary similar to the following:

```
> lmHeadToHeadRating_DurationSoldiers_Summary

Call:
lm(formula = subsetHeadToHead$Rating ~ subsetHeadToHead$DurationInDays +
    subsetHeadToHead$ShuSoldiersEngaged + subsetHeadToHead$WeiSoldiersEngaged,
    data = subsetHeadToHead)

Residuals:
     Min       1Q   Median       3Q      Max
-25.4560  -8.3716  -0.2642  10.0152  24.3812

Coefficients:
                                      Estimate Std. Error t value Pr(>|t|)
(Intercept)                         97.3500156  8.6961491  11.195 1.92e-11 ***
subsetHeadToHead$DurationInDays     -0.7680305  0.0958033  -8.017 1.70e-08 ***
subsetHeadToHead$ShuSoldiersEngaged  0.0005422  0.0001406   3.857 0.000678 ***
subsetHeadToHead$WeiSoldiersEngaged -0.0002755  0.0001049  -2.627 0.014258 *
---
Signif. codes:  0 '***' 0.001 '**' 0.01 '*' 0.05 '.' 0.1 ' ' 1

Residual standard error: 12.89 on 26 degrees of freedom
Multiple R-squared: 0.8585, Adjusted R-squared: 0.8421
F-statistic: 52.56 on 3 and 26 DF,  p-value: 3.57e-11
```

```
> #interaction model using duration, Shu soldiers, Wei
soldiers, and the interaction between Shu and Wei soldiers to
predict battle rating
> lmHeadToHeadRating_DurationSoldiersShuWeiInteraction_Summary
```

Produces the following summary:

```
>  lmHeadToHeadRating_DurationSoldiersShuWeiInteraction_Summary

Call:
lm(formula = subsetHeadToHead$Rating ~ subsetHeadToHead$DurationInDays +
    subsetHeadToHead$ShuSoldiersEngaged + subsetHeadToHead$WeiSoldiersEngaged +
    interactionSoldiersHeadToHead, data = subsetHeadToHead)

Residuals:
     Min       1Q   Median       3Q      Max
-25.3749  -8.9428   0.6675   9.3589  22.2165

Coefficients:
                                      Estimate Std. Error t value Pr(>|t|)
(Intercept)                          9.956e+01  8.871e+00  11.222 2.98e-11 ***
subsetHeadToHead$DurationInDays     -7.977e-01  9.890e-02  -8.066 2.02e-08 ***
subsetHeadToHead$ShuSoldiersEngaged  4.661e-04  1.553e-04   3.000  0.00603 **
subsetHeadToHead$WeiSoldiersEngaged -1.338e-04  1.635e-04  -0.819  0.42078
interactionSoldiersHeadToHead       -2.413e-09  2.143e-09  -1.126  0.27090
---
Signif. codes:  0 '***' 0.001 '**' 0.01 '*' 0.05 '.' 0.1 ' ' 1

Residual standard error: 12.83 on 25 degrees of freedom
Multiple R-squared: 0.8653, Adjusted R-squared: 0.8437
F-statistic: 40.14 on 4 and 25 DF,  p-value: 1.549e-10
```

2. Use `anova(object, ...)` to compare the relative contribution of each model:

```
> #use anova(object, ...) to compare the relative contribution
of multiple models
> #compare the three head to head combat models using ANOVA
> anovaHeadToHeadRatingModelComparison <-
anova(lmHeadToHeadRating_Duration,
lmHeadToHeadRating_DurationSoldiers,
lmHeadToHeadRating_DurationSoldiersShuWeiInteraction)
```

3. Display the `anova` results in the R console:

```
> display the anova results
> anovaHeadToHeadRatingModelComparison
```

```
> anovaHeadToHeadRatingModelComparison
Analysis of Variance Table

Model 1: subsetHeadToHead$Rating ~ subsetHeadToHead$DurationInDays
Model 2: subsetHeadToHead$Rating ~ subsetHeadToHead$DurationInDays
+ subsetHeadToHead$ShuSoldiersEngaged +
    subsetHeadToHead$WeiSoldiersEngaged
Model 3: subsetHeadToHead$Rating ~ subsetHeadToHead$DurationInDays
+ subsetHeadToHead$ShuSoldiersEngaged +
    subsetHeadToHead$WeiSoldiersEngaged +
interactionSoldiersHeadToHead
  Res.Df    RSS Df Sum of Sq      F   Pr(>F)
1     28 6964.9
2     26 4320.5  2    2644.4 8.0386 0.002015 **
3     25 4112.0  1     208.5 1.2676 0.270903
---
Signif. codes:  0 '***' 0.001 '**' 0.01 '*' 0.05 '.' 0.1 ' ' 1
```

What just happened?

You have the data that you need to complete a hierarchical linear regression (HLR) analysis. To be thorough, you should consider both the individual models (summaries) and the relative contribution of each model (ANOVA).

Interpreting the model summaries

You are already familiar with interpreting model summaries. These are the best places to start when conducting an HLR analysis. You can check the summaries to see if each overall model and its coefficients are statistically significant. You should also take note of each model's R-squared value.

Our simple regression model is statistically significant on all accounts and has an amiable R-squared value of 77%. Likewise, all of the variables in our multiple regression model, as well as the model itself, are statistically significant. The model has an R-squared value of 86%. Furthermore, while our interaction model is also statistically significant, with an R-squared of 87%, two of its predictor variables are not statistically significant. Although these summaries provide us with a wealth of knowledge on the individual merits of each model, it is best to make a decision after considering the results of an `anova` test.

Interpreting the ANOVA results

Generally, **analysis of variance (ANOVA)** is a statistical procedure that compares the means of multiple groups and determines if they are significantly different from one another. In our case, ANOVA can be used in HLR to compare multiple regression models. Here, ANOVA determines if the coefficient(s) that each successive model brings to the overall regression equation makes a statistically significant contribution above and beyond the coefficients that preceded it.

Consider the following three models:

```
A: Y = X1
B: Y = X1 + X2
C: Y = X1 + X2 + X3
```

The difference between each model is that a new predictor is contributed to the regression equation. Model *B* contributes *X2* in addition to model `A`, whereas model `C` contributes `X3` in addition to model `B`. ANOVA succeeds in determining whether these successive contributions are statistically significant. For instance, if model `B` was found to be statistically significant through ANOVA, then including `X2` in the regression model is likely to add value. Continuing, if model `C` were not found to be statistically significant, then including `X3` in the regression model probably does not add much value and therefore should be removed. By comparing successive models in this manner, we are able to determine, in a statistical sense, whether our coefficients are or are not adding value to the overall model. Thus, our decisions to include valuable coefficients and eliminate excess ones are informed.

Of course, we have to be mindful of practical significance at all times. When selecting independent variables for our model, we should use our understanding of the data and the situation to select only the best predictors. Although we could, it is inappropriate to haphazardly test numerous arbitrary combinations of variables in an attempt to find the supposed *best* statistical model. In fact, partaking in such practice is likely to lead to a model that is both meaningless in a practical sense and incapable of predicting valid answers to the questions that motivated the use of regression modeling in the first place. Therefore, always keep in mind the practical implications of every statistical analysis.

anova(object, ...)

R's `anova(object, ...)` is a variable-argument function that can be used to conduct ANOVA on several objects. Each object of comparison can be entered into the function as its own argument. For example, in:

```
anova(A, B, C)
```

Here we are telling R to compare three objects (`A`, `B`, and `C`) using ANOVA.

The `anova(object, ...)` function yields an **ANOVA table**, which details the results of the analysis. For the purposes of comparing successive models using HLR, we are only concerned with the p-values (the **Pr(>}|t|)** column). The p-value beside each model indicates whether or not it is statistically significant above and beyond its preceding model. It does not however, indicate the individual statistical significance of the model, which is why we also considered the individual model summaries.

```
> anovaHeadToHeadRatingModelComparison
Analysis of Variance Table

Model 1: subsetHeadToHead$Rating ~ subsetHeadToHead$DurationInDays
Model 2: subsetHeadToHead$Rating ~ subsetHeadToHead$DurationInDays
+ subsetHeadToHead$ShuSoldiersEngaged +
    subsetHeadToHead$WeiSoldiersEngaged
Model 3: subsetHeadToHead$Rating ~ subsetHeadToHead$DurationInDays
+ subsetHeadToHead$ShuSoldiersEngaged +
    subsetHeadToHead$WeiSoldiersEngaged +
interactionSoldiersHeadToHead
  Res.Df    RSS Df Sum of Sq      F   Pr(>F)
1     28 6964.9
2     26 4320.5  2    2644.4 8.0386 0.002015 **
3     25 4112.0  1     208.5 1.2676 0.270903
---
Signif. codes:  0 '***' 0.001 '**' 0.01 '*' 0.05 '.' 0.1 ' ' 1
```

The ANOVA table from our activity indicates that our multiple regression model is statistically significant above and beyond our simple regression model. However, our interaction model does not make a statistically significant contribution above and beyond our multiple regression model. This suggests, from a statistical standpoint, that our interaction coefficient should be removed. Recall that we did not formulate a logical basis for the interaction between the number of Shu and Wei soldiers engaged in head to head combat. Without a statistical or practical reason to include the interaction coefficient, it is best removed from the model. In other words, our HLR analysis suggests that, out of the models that we analyzed, the multiple regression model is best.

Pop quiz

1. Which of the following best explains the meaning of a statistically significant result in an ANOVA table generated during an HLR analysis?
   a. The regression models' coefficients are statistically significant.
   b. The overall regression model is statistically significant.
   c. The contribution that the model makes is statistically significant.
   d. The contribution that the model makes above and beyond the preceding model is statistically significant.

Have a go hero

Using the techniques that we explored in this chapter, analyze the remaining battle methods— surround, ambush, and fire— and create regression models for each that predict the performance rating of the Shu army. Be sure to use your practical knowledge of the combat strategies to choose appropriate coefficients for your regression models. Once you have found a few reasonably predictive models for each method, use HLR to compare them. Ultimately, come to a statistically and practically justifiable conclusion about the best regression model to use for each battle method. Remember to save your R workspace and console text to preserve the content that you created during this chapter.

Summary

Throughout this chapter, you explored your data for the purpose of weighing potential options. En route, you have considered both the practical and statistical significance of your decisions. You have derived four predictive regression models, one for each battle method, that you can use to develop and assess potential battle strategies for the Shu forces. At this point, you should be able to:

- Use multi-argument and variable-argument functions to make calculations
- Create predictive models using regression analysis
- Consider the statistical and practical significance of your analyses

Our next chapter will focus on using the models that we have developed, as well as our logistical constraints, to decide on an ultimate course of action for the Shu army.

6
Planning the Attack

In the preceding chapter, you developed four regression models to predict the outcomes of battles in which the Shu army uses head to head, surround, ambush, and fire attack methods. A sample regression model for each of the battle methods is provided to you in this chapter. For demonstration and consistency, these models will be used throughout the chapter. However, you are encouraged to substitute your own models from Chapter 5 into the calculations and activities in this chapter.

For the duration of this chapter, we will focus on employing our regression models to predict outcomes and to determine the feasibility of different attack strategies. Ultimately, you will need to decide on the best course of action for the Shu army. By the end of this chapter, you will be able to:

- Use regression models to predict outcomes
- Create your own custom functions to address specific needs
- Assess the viability of achieving the outcomes predicted by regression models

Review of models

In this section, we will review each of the four regression models created in *Chapter 5*. This will refresh our memory and prepare us to use our models in developing and assessing potential strategies. Again, while these sample models will appear throughout this chapter, feel free to substitute your own models into any or all activities.

Head to head

The following is a summary of the head to head model:

```
> modelHeadToHead_Summary

Call:
lm(formula = subsetHeadToHead$Rating ~ subsetHeadToHead$DurationInDays +
    subsetHeadToHead$ShuSoldiersEngaged + subsetHeadToHead$WeiSoldiersEngaged,
    data = subsetHeadToHead)

Residuals:
     Min       1Q   Median       3Q      Max
-25.4560  -8.3716  -0.2642  10.0152  24.3812

Coefficients:
                                      Estimate Std. Error t value Pr(>|t|)
(Intercept)                         97.3500156  8.6961491  11.195 1.92e-11 ***
subsetHeadToHead$DurationInDays     -0.7680305  0.0958033  -8.017 1.70e-08 ***
subsetHeadToHead$ShuSoldiersEngaged  0.0005422  0.0001406   3.857 0.000678 ***
subsetHeadToHead$WeiSoldiersEngaged -0.0002755  0.0001049  -2.627 0.014258 *
---
Signif. codes:  0 '***' 0.001 '**' 0.01 '*' 0.05 '.' 0.1 ' ' 1

Residual standard error: 12.89 on 26 degrees of freedom
Multiple R-squared: 0.8585, Adjusted R-squared: 0.8421
F-statistic: 52.56 on 3 and 26 DF,  p-value: 3.57e-11
```

Our head to head regression model predicts the Shu army's performance rating based on the duration of battle and the number of Shu and Wei soldiers engaged. All of these coefficients, as well as the overall model, are statistically significant. The model explains 86% of the variance in performance rating. Therefore, 14% of the rating remains unaccounted for and unpredicted. Our head to head regression equation is:

```
Rating = 97 -  0.77 * duration + 0.00054 * Shu soldiers - 0.00028
* Wei soldiers
```

Recall that our dependent variable of `Rating` is represented numerically on a scale from 0 to 100. Consequently, the higher the value predicted by our regression model, the more confident we can be that our strategy will lead to victory. Conversely, a lower value would make us more certain that our strategy would lead to defeat. For instance, a value of 90 would indicate a higher likelihood of victory, while a value of 10 would indicate a higher likelihood of defeat. Keeping this in mind, let us analyze the coefficients in our head to head combat model.

In our equation, the duration coefficient of -0.77 indicates that the Shu army's chances of victory decrease rapidly as the length of a head to head conflict increases. The positive coefficient for Shu soldiers engaged implies that deploying more Shu soldiers leads to a higher prospect of victory. In contrast, the negative coefficient for Wei soldiers engaged suggests that the more Wei soldiers deployed, the lower the chances of victory for the Shu army. The intercept of 97 does not have a logical practical interpretation, but it is essential to making predictions with the model. This is true of the intercept in each of our sample models.

Surround

The following is a summary of the surround model:

```
> modelSurround_Summary

Call:
lm(formula = subsetSurround$Rating ~ numericExecutionSurround +
    subsetSurround$DurationInDays + subsetSurround$ShuSoldiersEngaged +
    subsetSurround$WeiSoldiersEngaged, data = subsetSurround)

Residuals:
     Min       1Q   Median       3Q      Max
-11.6306  -3.2089  -0.4548   3.5928  15.4747

Coefficients:
                                    Estimate Std. Error t value Pr(>|t|)
(Intercept)                        3.470e+01  7.278e+00   4.768 6.80e-05 ***
numericExecutionSurround           5.765e+01  3.641e+00  15.832 1.54e-14 ***
subsetSurround$DurationInDays     -1.488e-01  5.900e-02  -2.522 0.018409 *
subsetSurround$ShuSoldiersEngaged  1.758e-04  3.817e-05   4.606 0.000104 ***
subsetSurround$WeiSoldiersEngaged -1.935e-04  5.587e-05  -3.463 0.001939 **
---
Signif. codes:  0 '***' 0.001 '**' 0.01 '*' 0.05 '.' 0.1 ' ' 1

Residual standard error: 5.769 on 25 degrees of freedom
Multiple R-squared: 0.9791, Adjusted R-squared: 0.9758
F-statistic: 293.4 on 4 and 25 DF,  p-value: < 2.2e-16
```

Our surround method regression model predicts the Shu army's performance rating based on execution (successful or unsuccessful), the duration of battle, and the number of Shu and Wei soldiers engaged. All of these coefficients, as well as the overall model, are statistically significant. This model contains a remarkable 98% of the elements that predict the variance in performance rating when the surround strategy is employed. Our surround regression equation is:

```
Rating = 35 + 58 * execution - 0.15 * duration + 0.18 *
Shu soldiers - 0.19 * Wei soldiers
```

Here, the 58 coefficient suggests that successful execution is not only critical, but likely necessary to predict victory. Recall that our `SuccessfullyExecuted` variable was categorical. It has been represented as 0 for no and 1 for yes. Accordingly, successful execution of the surround method will add 58 to our final rating prediction, whereas unsuccessful execution will contribute 0. Therefore, our predicted outcome weighs tremendously on whether or not we expect our forces to successfully execute the surround technique. Again, a shorter duration of battle is better. The coefficients for Shu and Wei soldiers engaged can be interpreted in similar fashion to our head to head model.

Ambush

The following is a summary of the ambush model:

```
> modelAmbush_Summary

Call:
lm(formula = subsetAmbush$Rating ~ numericExecutionAmbush + subsetAmbush$DurationInDays +
    subsetAmbush$ShuSoldiersEngaged + subsetAmbush$WeiSoldiersEngaged,
    data = subsetAmbush)

Residuals:
    Min      1Q  Median      3Q     Max
-20.019  -8.640   2.539   8.041  13.981

Coefficients:
                                  Estimate Std. Error t value Pr(>|t|)
(Intercept)                     55.7923922 11.3052112   4.935 4.41e-05 ***
numericExecutionAmbush          44.3273148  5.9280503   7.478 7.87e-08 ***
subsetAmbush$DurationInDays     -1.9748222  0.6454942  -3.059 0.005233 **
subsetAmbush$ShuSoldiersEngaged  0.0017928  0.0004779   3.752 0.000935 ***
subsetAmbush$WeiSoldiersEngaged -0.0008191  0.0002641  -3.101 0.004731 **
---
Signif. codes:  0 '***' 0.001 '**' 0.01 '*' 0.05 '.' 0.1 ' ' 1

Residual standard error: 10.53 on 25 degrees of freedom
Multiple R-squared: 0.9165, Adjusted R-squared: 0.9031
F-statistic: 68.57 on 4 and 25 DF,  p-value: 4.154e-13
```

Our ambush method regression model predicts the Shu army's performance rating based on execution, duration, and the number of Shu and Wei soldiers engaged. All of these coefficients, as well as the overall model, are statistically significant. This model explains a formidable 92% of the variance in performance rating when the ambush strategy is employed. Our ambush regression equation is:

```
Rating = 56 + 44 * execution - 1.97 * duration + 0.0018 *
Shu soldiers - 0.00082 * Wei soldiers
```

In this case, the rating prediction is also tied strongly to successful execution. Once again, the duration and number of Shu and Wei soldiers engaged can be interpreted in the same manner as our preceding models.

Fire

The following is a summary of the fire model:

```
> modelFire_Summary

Call:
lm(formula = subsetFire$Rating ~ numericExecutionFire + subsetFire$DurationInDays +
    interactionSoldiersFire, data = subsetFire)

Residuals:
    Min      1Q  Median      3Q     Max
-18.233  -7.248   1.466   6.452  10.535

Coefficients:
                           Estimate Std. Error t value Pr(>|t|)
(Intercept)               3.737e+01  3.467e+00  10.780 4.34e-11 ***
numericExecutionFire      5.602e+01  3.486e+00  16.071 5.08e-15 ***
subsetFire$DurationInDays -1.237e+00  3.960e-01  -3.125  0.00434 **
interactionSoldiersFire   -1.273e-07  3.717e-08  -3.424  0.00206 **
---
Signif. codes:  0 '***' 0.001 '**' 0.01 '*' 0.05 '.' 0.1 ' ' 1

Residual standard error: 8.386 on 26 degrees of freedom
Multiple R-squared: 0.928,  Adjusted R-squared: 0.9197
F-statistic: 111.7 on 3 and 26 DF,  p-value: 5.638e-15
```

Our fire attack regression model predicts the Shu army's performance rating based on execution, duration, and the interaction between the number of Shu and Wei soldiers engaged in battle. Here, it is not the raw number of soldiers for each side that impacts our prediction, but rather the relationship between them. All of the coefficients, as well as the overall model, are statistically significant. This model explains a solid 93% of the variance in performance rating when the fire attack strategy is employed. Our fire attack regression equation is:

```
Rating = 37 + 56 * execution - 1.24 * duration - 0.00000013 *
soldiers interaction
```

In this equation, successful execution plays a critical role in explaining the battle rating, as does duration. Our interaction term suggests that the more soldiers involved in the battle, regardless of affiliation, the less likely our fire attack is to lead to victory. This makes sense considering that the fire attack, unlike our other methods, is a risky surprise tactic. Having too many Shu soldiers increases the visibility of our attack and the likelihood that our plans would be discovered. A similar condition arises from launching a fire attack on too many Wei soldiers. There would be more eyes to discover and arms to quash the surprise attack. Therefore, the interaction between the number of Shu and Wei soldiers involved in a fire attack must be balanced to optimize our impact and chances of success.

Predicting outcomes using regression models

Having reviewed each of our models, let us now look at how to use them to predict outcomes in R. Before we do so, we must address a few assumptions about our models.

Rating

In order to decide whether a strategy is sufficient or not, we must determine an acceptable `Rating` value. Assume for the remainder of this book that we consider a `Rating` value of 80 to be sufficient for predicting victory. After all, Zhuge Liang's rating of the Shu army's performance in each victorious campaign was 80 or higher. As such, a `Rating` of 80 or above makes us reasonably confident that our strategy will lead to victory. A `Rating` below 80 will be considered too risky and should be avoided.

Feel free to experiment with your own `Rating` values. The higher the threshold, the more certain you can be of victory, but the less likely you are to have the resources to achieve it. The lower the required value, the more resource allocation options you will have, but the higher risk your strategy has of failure.

Successfully executed

The outcomes of our surround, ambush, and fire attack regression models depend to a large extent on the successful execution of these battle methods. Yet, successful execution is not something that we can predict with certainty before a battle takes place. One way to handle this conundrum is to use our past battle data to calculate the probability that our battle methods will be successfully executed. Once obtained, we can enter our probability value into our regression equations to make more accurate predictions than we would by merely assuming that our methods were successfully or unsuccessfully executed.

To calculate our probability values, we need to look at the number of times that our methods were successfully executed in the past and divide them by the total number of battles that we have on record. For instance, we know that the Shu army successfully executed a fire attack in 10 out of 30 battles. Therefore, our probability value for successful execution of the fire attack method would be 0.33 (10 divided by 30). Identical steps can be taken to derive probability values for each of the battle methods. These values are displayed in the following table:

Method	Variable Name	Probability
Head to head	`probabilitySuccessHeadToHead`	1.00
Surround	`probabilitySuccessSurround`	0.53
Ambush	`probabilitySuccessAmbush`	0.50
Fire	`probabilitySuccessFire`	0.33

We will use these probability values for the `SuccessfullyExecuted` variable when making predictions with our regression models. However, do not hesitate to experiment with hypothetical scenarios. For instance, suppose you feel that your soldiers are better trained today than they have ever been in the past. Perhaps then they are more likely to successfully execute battle plans and deserve a higher probability value.

Number of Wei soldiers

Similarly, we cannot always determine how many soldiers the opposing army will bring into a given battle. However, we do have the history of 120 prior battles that can give us an idea of how many soldiers the Wei army tends to engage in relation to our own. From our past data, we can predict the ratio of Wei soldiers to Shu soldiers for each method by summing the total soldiers engaged for both sides and then dividing them. For example, in previous battles where the surround method was employed, the Wei army engaged a total of 820,000 soldiers, while the Shu army engaged 1,287,000. Accordingly, the ratio of Wei to Shu soldiers engaged was 0.64 (820,000 divided by 1,287,000). When the number of Wei soldiers is unknown, we will assume the following ratios of Wei to Shu soldiers for each type of battle:

Method	Variable name	Ratio
Head to head	`ratioWeiShuSoldiersHeadToHead`	1.08
Surround	`ratioWeiShuSoldiersSurround`	0.64
Ambush	`ratioWeiShuSoldiersAmbush`	1.82
Fire	`ratioWeiShuSoldiersFire`	6.01

Yet at times, we can indeed predict the number of soldiers that the Wei army will engage in battle. Imagine an ambush attack where we specifically target an enemy city with a known number of soldiers. In this case, it would be better to use the known value than the ratio estimate. When devising your final strategy, the appropriate assumptions will be determined by the particular situation at hand.

Duration of battle

Yet again, the duration of battle is something that is predictable in some cases and unpredictable in others. Our past data show that, on average, surround attacks last for a relatively long time, whereas fire attacks tend to be brief. We can derive the average duration of battle for each of our combat methods for use in instances where a battle's length cannot be predicted beforehand.

Method	Variable name	Average
Head to head	`meanDurationHeadToHead`	77.9
Surround	`meanDurationSurround`	105.5
Ambush	`meanDurationAmbush`	13.6
Fire	`meanDurationFire`	6.9

Whenever possible, it is best to determine the duration of battle. For example, when issuing a fire attack, you may explicitly set a small window of time in which the army is expected to execute its plans. Should the mission not be completed in time, it may be aborted, thereby remaining within the determined time frame.

A word about assumptions

As noted, these assumptions are not set in stone. In fact, they are our best effort to make the most reasoned and valid predictions in a situation where we cannot control or determine every possible variable. As you progress through this chapter and begin making your own battle predictions and plans, you may want to alter these assumptions to better fit your interpretation of a situation. Along the way, always remain conscious that the assumptions that you choose have important implications for the validity and riskiness of your predictions.

Time for action – calculating outcomes from regression models

With our necessary assumptions decided upon, the simplest way to predict the outcome of a particular scenario is to plug relevant numbers into a regression model and calculate the result. Suppose we want to estimate the rating of an ambush attack when we know only the number of Shu soldiers that we will engage.

1. Open R and set your working directory, as follows:

```
> #set the R working directory
> #replace the sample location with one that is relevant to you
> setwd("/Users/johnmquick/rBeginnersGuide/")
```

2. Load the *Chapter 6* workspace. It contains the sample models, as well as our assumed variables:

```
> #load the chapter six workspace
> load("rBeginnersGuide_Ch_06_ReadersCopy.RData")
```

3. Calculate the rating of an ambush attack in which the Shu forces engage 5,000 soldiers:

```
> #what is the predicted rating of an ambush attack in which
the Shu forces engage 5000 soldiers?
>#ambush model: Rating = 56 + 44 * execution - 1.97 * duration
+ 0.0018 * Shu soldiers - 0.00082 * Wei soldiers
>  56 + 44 * probabilitySuccessAmbush - 1.97 *
meanDurationAmbush + 0.0018 * 5000 - 0.00082 * (5000 *
ratioWeiShuSoldiersAmbush)
[1] 52.746
```

Our calculated rating of 53 falls well below our threshold of 80 and suggests that the Shu army would have roughly an equal chance of winning or losing in this battle scenario.

What just happened?

We just employed the simplest, but most time consuming, method for predicting outcomes from our regression models. Let us make this process more efficient by creating a custom function in R that automatically calculates a solution when we provide the known values.

Time for action – creating custom functions

In R, `function()` can be used to define a custom function, along with its arguments. This allows us to extend the capabilities of R by creating functions that meet our specific needs.

1. Use `function()` to define a custom function with arguments:

```
> #use function() to define custom function
> #define our ambush regression model as a custom function in R
> functionAmbushRating <- function(execution, duration,
ShuSoldiers, WeiSoldiers) {
+                             56 + 44 * execution -
+                             1.97 * duration +
+                             0.0018 * ShuSoldiers - 0.00082 *
+                             WeiSoldiers
+                             }
```

2. Test and verify the function:

```
> #what is the predicted rating of an ambush attack in which
the Shu forces engage 5000 soldiers?
> functionAmbushRating(probabilitySuccessAmbush,
meanDurationAmbush, 5000, 5000 * ratioWeiShuSoldiersAmbush)
[1] 52.746
```

   As you can see, our custom function resulted in the same value as our previous calculation. Conveniently, deriving this rating only required us to input the variable values, rather than solving each piece of the equation mathematically.

3. To further demonstrate our function, suppose instead that our 5,000 Shu soldiers are going to ambush a vulnerable unit of 1,000 Wei soldiers. Calculate the rating of an ambush attack by 5,000 Shu soldiers against 1,000 Wei soldiers:

```
> #what is the predicted rating of an ambush attack by 5000 Shu
soldiers against 1000 Wei soldiers?
> functionAmbushRating(probabilitySuccessAmbush,
meanDurationAmbush, 5000, 1000)
[1] 59.388
```

   Under a more predictable and favorable circumstance, our `Rating` value increased a small amount to 59. Here, we are a little more confident in victory than in our previous scenario, but still far from comfortable.

4. For a final test, let us examine the performance rating if we are completely certain that our forces will successfully execute the proposed ambush attack. Calculate the rating of a successful ambush attack by 5,000 Shu soldiers against 1,000 Wei soldiers:

```
> #what is the predicted rating of a successful ambush attack
by 5000 Shu soldiers against 1000 Wei soldiers?
> functionAmbushRating(1, meanDurationAmbush, 5000, 1000)
[1] 81.388
```

At 81, we are feeling pretty good about our chances for victory. But we must ask ourselves just how likely the proposed circumstances are to occur in an authentic battle situation. Naturally, our prediction is only valid to the extent that we believe that our estimates will reflect actual battle conditions.

What just happened?

We just explored the creation and use of custom functions in R. The ability to create custom functions is a powerful feature that allows you to expand the capabilities of the software to meet your personal needs. Let us discuss the details of custom functions.

function()

In R, the `function()` command can be used to create custom functions. These can take many shapes and forms. They can also have anywhere from zero to several arguments. The basic format of the `function()` command is as follows:

```
function(argument1, argument2,... argumenti) { contents }
```

Here are some examples of custom functions:

- No arguments:

  ```
  function() { setwd("/Users/johnmquick/Desktop") }
  ```

  This function sets the working directory to the desktop.

- One argument:

  ```
  function(path) { setwd(path) }
  ```

  This function sets the working directory to a specified path.

- Multiple arguments:

  ```
  function(path, verify) {
                              setwd(path)
                              if (verify == TRUE) {
                                                          getwd()
                                                          }
                              }
  ```

  This function sets the working directory to a specified path and then optionally reports that path in the R console.

As we demonstrated in the preceding activity, it is often useful to save a custom function into an R variable. This saves you the effort of retyping the entire command each time you want to execute the function. Furthermore, it allows you to call the function, complete with arguments, using the variable name. These benefits are demonstrated in the following sample code:

```
> #save a custom function into an R variable
> customFunction <- function(x,y) { 5 * x + 2 * y }
> #call the function by its variable name and solve for x = 1 and
y = 2
> customFunction(1,2)
[1] 9
```

Note that the parenthesis () are required when you want to execute a function that has been saved into a variable. Without them, the contents of the variable will be displayed in the R console. These differences are demonstrated in the following:

```
> #without parenthesis, the contents of the function are displayed
> customFunction
function(x,y) { 5 * x + 2 * y }

> #with parenthesis, the function is executed
> customFunction(1,2)
[1] 9
```

Extended lines

When we created our custom function in step 1 of the previous activity, you may have noticed a new type of console line. These **extended lines** begin with a plus (+) sign. Unlike input lines that begin with a greater than sign (>) and output lines that have no leading character, extended lines are purely cosmetic. Extended lines are used to format long segments of code so that they are more readable and aesthetically pleasing. The plus sign lets you know that your previous line is being continued. In effect, an extended line is similar to using a hard return in a word processor. The previous line is cut off immediately, while the text continues at the start of the next line. The formatting value of extended lines is clarified by the following sample code:

```
> #using a single line to define a long function
> functionAmbushRating <- function(execution, duration,
ShuSoldiers, WeiSoldiers) { 56 + 44 * success - 1.97 * duration +
0.0018 * ShuSoldiers - 0.00082 * WeiSoldiers }

> #using multiple lines to define a long function
> functionAmbushRating <- function(execution, duration,
ShuSoldiers, WeiSoldiers) {
+                          56 + 44 * execution -
+                          1.97 * duration +
+                          0.0018 * ShuSoldiers - 0.00082 *
+                          WeiSoldiers
+                          }
```

Pop quiz

1. Which of the following elements is not required when creating a custom function?

   `function(argument`$_1$`, argument`$_2$`,... argument`$_i$`) { contents }`

   a. `function`

   b. `()`

   c. `argument`$_1$`, argument`$_2$`,... argument`$_i$

   d. `contents`

2. Which of the following is not true of a variable that contains a custom function?

   a. It can be redefined to store a new custom function or other data.

   b. Its function can be called by typing the variable name in the R console.

   c. Its contents can be displayed by typing the variable name in the R console.

   d. Its function can be called by typing the variable name, along with the function's arguments in parenthesis, in the R console.

3. What does a plus sign (+) at the beginning of an R console line indicate?

   a. The mathematical addition operator.

   b. A line of code that is contained within a custom function.

   c. A single line of code that is being extended across multiple console lines.

   d. Multiple lines of code that are being extended across multiple console lines.

Have a go hero

Now that you are familiar with generating custom functions, use the `function()` command to recreate the regression equations for each of the remaining battle methods—head to head, surround, and fire—as R functions. Save each of these custom functions into new variables, named `functionHeadToHeadRating`, `functionSurroundRating`, and `function0FireRating` respectively. Then test each of your functions using the hypothetical battle data.

Time for action – creating resource-focused custom functions

Rather than plugging in values to calculate the outcome of a specific scenario, suppose that we instead choose to determine the resources necessary to realize a desired result. In other words, a different way to approach the challenge of developing a successful battle plan is to set our required outcome, say a `Rating` of 80, and then solve for the number of soldiers or other resources needed to achieve that outcome. We can make this process possible through the use of custom functions.

The following procedure describes how to determine the amount of resources needed to achieve a `Rating` of 80 in our ambush regression model:

1. Solve the regression equation for the variable of interest:

```
> #rearrange the ambush model to solve for the number of Shu
soldiers engaged
> #original ambush model: Rating = 56 + 44 * execution - 1.97 *
duration + 0.0018 * Shu soldiers - 0.00082 * Wei soldiers
> #ambush model solved for Shu soldiers: (Rating - 56 - 44 *
execution + 1.97 * duration + .0.00082 * Wei soldiers) / 0.0018
```

2. Create a custom function for the rearranged model and save it into an R variable:

```
> #convert the rearranged ambush model equation into a custom
function
> functionAmbushShuSoldiers <- function(rating, execution,
duration, WeiSoldiers) {
+                        (rating - 56 - 44 * execution +
+                        1.97 * duration +
+                        0.00082 * WeiSoldiers) /
+                        0.0018
+                        }
```

3. Test the function:

```
>  #how many Shu soldiers must be engaged in an ambush attack
against 10,000 Wei soldiers to bring our rating to 80?
>  functionAmbushShuSoldiers(80, probabilitySuccessAmbush,
meanDurationAmbush, 10000)
[1] 20551.11
>  #how many Shu soldiers must be engaged in an ambush attack
against 10,000 Wei soldiers to bring our rating to 80 if we are
certain of successful execution?
>  functionAmbushShuSoldiers(80, 1, meanDurationAmbush, 10000)
[1] 8328.889
```

Each of our regression equations can be rearranged in the same manner as our ambush model. By solving for the number of Shu soldiers in our combat models, we can calculate the amount of resources that our army must expend in specific situations. This approach allows us to focus on determining the amount of resources required to achieve our desired outcomes.

What just happened?

We again employed the `function()` command to create a custom function based on one of our regression models. This activity represented a resource-focused approach to predicting the outcomes of potential battle situations.

Have a go hero

Use the `function()` command to create resource-focused custom functions for each of the remaining battle methods—head to head, surround, and fire. Save these custom functions into new R variables, named appropriately for the data variable that you focused on. For example, our ambush function in the previous activity solved for the number of Shu soldiers engaged and thus was named `functionAmbushShuSoldiers`. Afterwards, test each of your functions using hypothetical battle data.

Logistical considerations

Up to this point, we have paid little attention to resource constraints. Instead, we explored the range of possibilities to obtain the most optimal prediction models and outcomes. Yet, as the time to make a decision rapidly approaches, we must pay heed to the practical limitations of the Shu army. When considered within the context of our predictive models, logistical constraints will reveal our realistic set of available opportunities. From these, we can determine the risks and rewards of our potential actions with increased confidence. Taking our resources into account will lead us towards a sound and relevant decision.

The following sections outline the specific resource allotments available to the Shu army and the costs associated with the current campaign. Since we will use this information throughout the chapter, you may want to bookmark this segment for future reference. We will use four resources to determine the viability of our battle plans. These are gold, provisions, equipment, and soldiers.

Gold

Gold is our form of currency. Most actions have some kind of monetary cost associated with them. The emperor has allotted the army *1,000,000 gold*. This can be used however you wish in devising a strategy for the Shu forces.

Provisions

Provisions are needed to sustain the human component of the Shu army. You have *1,000,000 in provisions* available. The amount required per month (30 days) depends on the number of soldiers that you take into battle. Therefore, more soldiers equates to a faster rate of consumption. Since each soldier needs one unit of provisions per 30 days to survive, *the daily consumption rate for the Shu army is equal to the number of soldiers engaged in battle divided by 30* and *the cost of purchasing new provisions is one gold per unit*. Thus, an army of 100,000 soldiers would require 100,000 provisions to sustain itself for 30 days (`100,000 soldiers * (30 / 30) = 100,000 provisions`).

Equipment

Equipment refers to the manufactured component of the Shu army. This includes items such as weapons, armor, chariots, tents, and so on. The army's *equipment depreciates at a rate of 0.1 (10%) per month (30 days)*. You need to pay *one gold per soldier* engaged in battle to keep your equipment in prime condition. For instance, the equipment upkeep for an army of 100,000 would cost 100,000 gold per 30 days. Without maintained equipment, the Shu forces stand no chance against the Wei, who have greater human and manufactured resources.

Soldiers

Soldiers are the human resource of the Shu army. The number of soldiers that you take into battle has a tremendous impact on the expenses incurred, as well as the outcome of the conflict. Therefore, it is a matter of delicate balance. Hanzhong, the site from which you will launch your upcoming attack, currently has *100,000* soldiers. You have the option to relocate soldiers to different cities within your kingdom. All things considered, the *cost to move soldiers between cities is one gold per soldier per 100 miles*. Keep in mind that strengthening one location is equivalent to weakening another. It is best to make sure that all critical cities within your kingdom are sufficiently staffed to protect against invasion. The resource map (see the *Resource map* section of this chapter) depicts the current distribution of soldiers in various cities within the Shu and Wei kingdoms. From this, we can see that moving 1,000 soldiers from Baxi to Hanzhong, a distance of 100 miles, would cost 1,000 gold.

Resource and cost summary

The following table summarizes the logistical considerations that you will need to attend to while devising a strategy for the Shu forces:

Resource	Quantity	Cost
Gold	1,000,000	
Provisions	1,000,000	1 gold per soldier per 30 days to sustain
Equipment	1.0	1 gold per soldier per 30 days to maintain
Soldiers	100,000	1 gold per soldier per 100 miles to relocate

With these resources and costs in mind, you can predict the outcomes and assess the feasibility of potential battle plans.

Resource map

The following map details the locations of cities and the distribution of soldiers within the Shu and Wei kingdoms. You should use this information to predict outcomes and determine the feasibility of your proposed strategies.

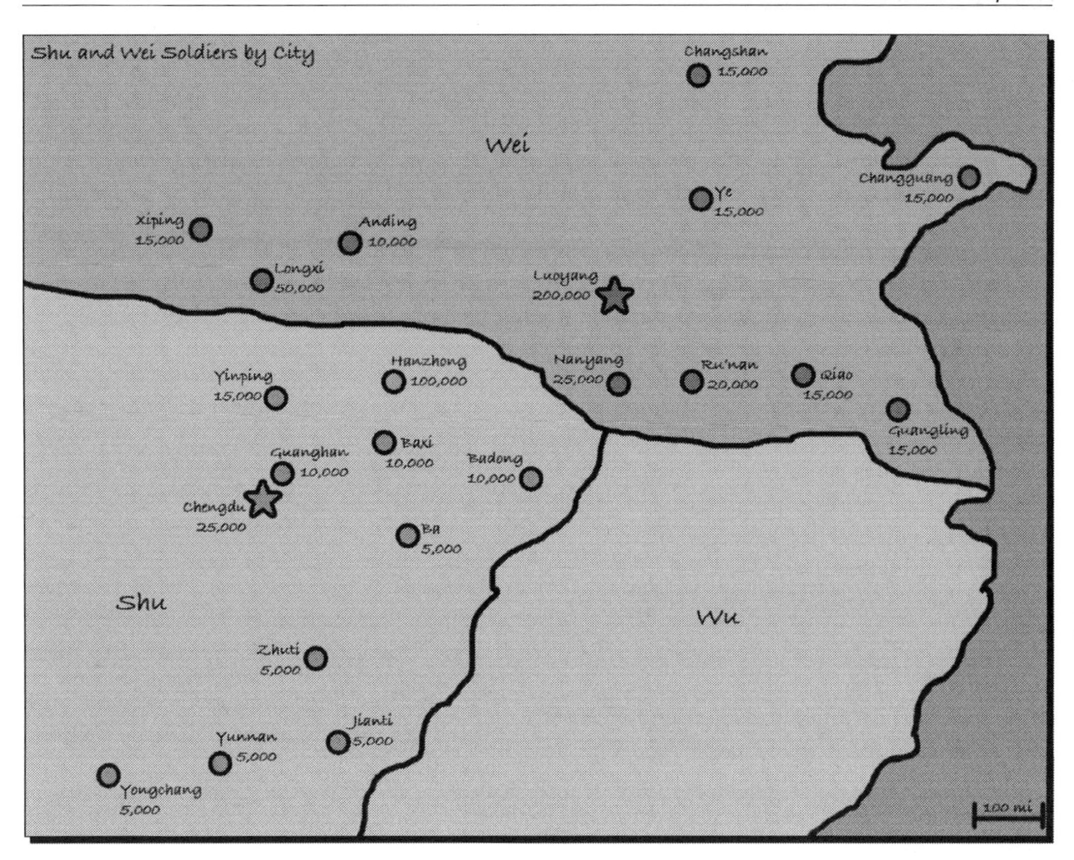

Time for action – incorporating resource constraints into predictions

Since we have added resource constraints as an additional factor in our decision process, let us create a custom function to calculate the gold cost for a given battle plan:

1. Define a custom function that calculates the gold cost of a battle when its distance, duration, and number of Shu soldiers engaged are known:

```
> #custom function that calculates the gold cost of a battle
> #cost formula: travel cost + (provision cost +
equipment cost) * battle duration
> functionGoldCost <- function(ShuSoldiers, distance, duration)
+       {
+        ShuSoldiers * ((distance / 100) + 2 * (duration / 30))
+        }
```

2. Test the function:

```
> #what is the predicted cost of an attack by 25,000 Shu
soldiers that takes place 100 miles away and lasts for 90 days?
> functionGoldCost(25000, 100, 90)
[1] 175000
```

We now have a way to calculate the gold cost of our potential strategies. Alternatively, you could also choose to create functions solving for other combat constraints, such as soldiers, distance, or duration.

What just happened?

We created a custom function that tells us how much gold we would need to execute our plans when the number of Shu soldiers, the distance to the attack site, and the proposed duration of the battle are known.

Gold cost function explanation

The formula that we used in our gold cost function may seem unfamiliar. This is because it was coded in its simplest, and therefore easiest to read, form. The expanded formula for calculating our gold cost is detailed as follows:

```
ShuSoldiers * distance / 100 + ShuSoldiers * provision cost *
duration + ShuSoldiers * equipment cost * duration
```

Once simplified, we are left with the formula used in our gold cost function:

```
ShuSoldiers * ((distance / 100) + 2 * (duration / 30))
```

The `ShuSoldiers` term has been extracted and placed at the front of the equation. The `distance` is divided by 100, because the cost of moving one soldier is one gold per 100 miles. The `duration` is multiplied by two and divided by 30, because the cost of provisions and equipment are both one gold per soldier per 30 days of battle. In the end, we have the same output as with our expanded formula, but using much less space.

Pop quiz

1. Which of the following is **not** a reason to carefully consider the logistics of predicted outcomes?
   a. Considering logistics helps us to account for resource constraints.
   b. Considering logistics helps us to identify a realistic set of opportunities.
   c. Predicted outcomes are not always logistically viable.
   d. Predicted outcomes present the most logistically sound course of action.

Have a go hero

Create a custom function that tells us how many miles our army can travel given the proposed amount of gold, number of soldiers, and duration of the attack. Save it into a variable named `functionMaxDistance`. This function will prove useful in assessing the viability of the strategies predicted by our regression models.

Assessing viability

Our last major step in choosing a battle strategy is to assess its viability in light of the resource constraints imposed upon our forces. We can accomplish this by taking our best plans, calculating the costs associated with executing them, and then comparing those costs to the resources that we have available.

For the time being, suppose we have decided to explore the possibilities of a 7-day fire attack on the 10,000 strong Wei army 255 miles away at Anding. Also assume that we have already predicted the conditions necessary for a rating of 80, as demonstrated:

```
> #custom function that calculates the number of Shu soldiers
needed to execute a given fire attack
> functionFireShuSoldiers <- function(rating, execution, duration,
WeiSoldiers) {
+           (rating - 37 - 56 * execution +
+           1.24 * duration) /
+           (0.00000013 * - WeiSoldiers)
+           }
> #assuming successful execution, how many Shu soldiers would be
needed to lead a fire attack against the 10,000 Wei soldiers
stationed 225 miles away at Anding?
> functionFireShuSoldiers(80, 1, 7, 15000)
[1] 2215.385
```

Through this process, we determined that 2,215 Shu soldiers must successfully launch the 7-day fire attack to meet our `Rating` threshold of 80.

Our next step is to take this information and examine its viability in terms of the resources that have been allotted to us.

Time for action – assessing the viability of potential strategies

Thus far, we have looked at two ways to factor resources into our predictions. One focuses on calculating the gold cost of a mission, whereas the other searches for the maximum distance over which our proposed campaign could take place. We will demonstrate both approaches here:

1. Calculate the gold cost of the planned mission:

```
> #how much would it cost to launch a 7 day fire attack with
2,215 soldiers, 225 miles away in Anding?
> functionGoldCost(2215, 225, 7)
[1] 6017.417
```

2. Calculate the distance over which the proposed attack could take place:

```
> #custom function that calculates the the maximum distance the
Shu army can travel given our resources
> functionMaxDistance <- function(gold, ShuSoldiers, duration)
+           {
+           100 * ((gold / ShuSoldiers) + (duration / 15))
+           }
> #how many miles can a Shu force of 2215 travel to execute a 7
day fire attack, given our allotment of 1,000,000 gold?
> functionMaxDistance(1000000, 2215, 7)
[1] 45193.39
```

In our first calculation, we derived a cost of 6,017 gold for the attack. This is well under our budget of 1,000,000 and therefore is completely practical. Our second calculation found the maximum possible distance that our army could travel, given our resources. The distance of 45,193 miles is well beyond the distance to the target city. Therefore, we have also determined the distance of our attack to be feasible.

What just happened?

We looked at assessing the viability of a potential fire attack in terms of gold cost and travel distance. These are just a pair of the numerous possible ways that the practicality of our plans could be tested and confirmed. Be sure to explore every angle necessary to make yourself confident that your plans are the best ones available. After all, the welfare of many people depends upon your decisions.

Remember your assumptions

One final reminder is that we must be wary of the assumptions that we make in formulating plans. If you look back at our calculation that lead to a requirement of 2,215 soldiers, you will notice that we assumed our fire attack would be successful. We must ask ourselves if this, as well as any, assumption is a sound one.

According to our past battle data, fire attacks have only been executed successfully 33% of the time. Let us look at how our viability would change if we were to use this probability value, rather than assuming total success:

```
> #based on past battle data, how many Shu soldiers would be
needed to lead a victorious 7 day fire attack against the 10,000
Wei soldiers at Anding?
> functionFireShuSoldiers(80, probabilitySuccessFire, 7, 10000)
[1] -25538.46
```

Our recommended number of soldiers has suddenly turned negative! Considering that engaging a negative number of soldiers is an obvious impossibility, this indicates that our fire attack plans are completely impractical. This example has demonstrated how changing one simple assumption can have a dramatic impact on our predictions and subsequent decisions.

You may be wondering which assumption, 1.0 or 0.33, is the better one in our case. As with all assumptions, the truth is that there is no absolute answer. Since our work deals with predicting the future, there will always be uncertainties about the assumptions that we make. The best that we can do is to thoughtfully consider all of the information available to us. In doing so, we can derive predictions that most accurately reflect the conditions present in the world.

Pop quiz

1. Which of the following is **not** a reason to carefully consider assumptions when making logistical considerations?
   a. Assumptions rarely have an absolute best answer.
   b. Assumptions often have an impact on calculated results.
   c. Assumptions may affect the validity of predicted outcomes.
   d. Assumptions can be altered to achieve desirable real-world results.

Have a go hero—choosing a battle plan

You have worked long and hard to learn the techniques of master strategist Zhuge Liang. Furthermore, you have become deeply aware of and involved in the circumstances surrounding the Shu army. The time has come for you to determine which course of action the Shu forces will take. Use the knowledge and skills that you have acquired throughout this journey to predict and assess the optimal strategy for your army's upcoming attack. It is recommended that you explore all four methods available—head to head, surround, ambush, and fire—before making a final decision.

The following table has been provided to aid you in this process. One scenario has been calculated in the head to head column as an example. By the end of this activity, you should decide on the strategy that your forces will execute in the upcoming battle:

Potential battle strategies

Method	Head to head	Surround	Ambush	Fire
Shu Soldiers	87,376			
Wei Soldiers	15,000			
Predicted Execution	1.0			
Predicted Rating	80			
Location of Attack	Guangling			
Distance	700			
Duration	78			
Provisions	227,178			
Gold Cost	1,065,987			
Viable	No			

Summary

During this chapter, you used several custom functions to predict outcomes and then evaluated your predictions from a practical perspective. Ultimately, you determined the best strategy available for the Shu army's next attack. While coming to this conclusion, you acquired the knowledge and skills necessary to:

- Use regression models to predict outcomes
- Create your own custom functions to address specific needs
- Assess the viability of achieving the outcomes predicted by regression models

While you may have decided on a course of action for the Shu forces, your job is far from over. The major challenge at this point is to convey your ideas to others in such a way that they can be easily understood.

The next section of our book deals with the challenge of communicating the results of our data analyses. In *Chapter 7*, we will focus on conducting a complete, organized analysis in R. In *Chapter 8*, we will seek support from the emperor by presenting our battle plans graphically. In *Chapter 9*, we will use detailed custom visuals to educate our generals on the strategy that they will execute.

7

Organizing the Battle Plans

In the previous chapter, you completed your data analysis and selected the optimal course of action for the Shu army. With this decision in place, the time has come for you to share your strategy with the Shu forces. The initial step towards communicating your vision to the masses is to revisit the work that you have done up to this point.

In this chapter, we will focus on reorganizing and clarifying our prior analyses such that they can be easily followed by and communicated to others. This will render our work intelligible to a large audience, even if it is composed of members who do not have the exceptional level of expertise in data analysis, military strategy, statistics, and R that you do. Along the way, you will learn the common steps that you can apply to all of your future analyses in R.

By the end of this chapter, you will be able to:

- Organize and clarify your raw R data analyses
- Communicate your raw R data analyses effectively
- Apply the steps common to all well-conducted R analyses

Retracing and refining a complete analysis

For demonstration purposes, it will be assumed that a fire attack was chosen as the optimal battle strategy. Throughout this segment, we will retrace the steps that lead us to this decision. Meanwhile, we will make sure to organize and clarify our analyses so they can be easily communicated to others.

> Note that at the end of this chapter, you will be challenged to repeat these steps with the strategy that you devised in *Chapter 6*. However, if you feel comfortable using your own battle plans from the start, you are encouraged to do so.

Suppose we determined our fire attack will take place 225 miles away in Anding, which houses 10,000 Wei soldiers. We will deploy 2,500 soldiers for a period of 7 days and assume that they are able to successfully execute the plans. Let us return to the beginning to develop this strategy with R in a clear and concise manner.

Time for action – first steps

To begin our analysis, we must first launch R and set our working directory:

1. Launch R.
2. The R console will be displayed.
3. Set your R working directory using the `setwd(dir)` function. The following code is a hypothetical example. Your working directory should be a relevant location on your own computer.

   ```
   > #set the R working directory using setwd(dir)
   > setwd("/Users/johnmquick/rBeginnersGuide/")
   ```

4. Verify that your working directory has been set to the proper location using the `getwd()` command:

   ```
   > #verify the location of your working directory
   > getwd()
   [1] "/Users/johnmquick/rBeginnersGuide/"
   ```

What just happened?

We prepared R to begin our analysis by launching the software and setting our working directory. At this point, you should be very comfortable completing these steps.

Time for action – data setup

Next, we need to import our battle data into R and isolate the portion pertaining to past fire attacks:

1. Copy the `battleHistory.csv` file into your R working directory. This file contains data from 120 previous battles between the Shu and Wei forces.

2. Read the contents of `battleHistory.csv` into an R variable named `battleHistory` using the `read.table(...)` command:

```
> #read the contents of battleHistory.csv into an R variable
> #battleHistory contains data from 120 previous battles
between the Shu and Wei forces
> battleHistory <- read.table("battleHistory.csv", TRUE, ",")
```

3. Create a subset using the `subset(data, ...)` function and save it to a new variable named `subsetFire`:

```
> #use the subset(data, ...) function to create a subset of
the battleHistory dataset that contains data only from battles
in which the fire attack strategy was employed
> subsetFire <- subset(battleHistory, battleHistory$Method ==
"fire")
```

4. Verify the contents of the new subset. Note that the console should return 30 rows, all of which contain `fire` in the `Method` column:

```
> #display the fire attack data subset
> subsetFire
```

```
> subsetFire
    Method Rating SuccessfullyExecuted  Result ShuSoldiersEngaged WeiSoldiersEngaged DurationInDays
91    fire     30                    N  Defeat                100               1500              4
92    fire    100                    Y Victory                100               1500              1
93    fire     25                    N  Defeat                100               2500              2
94    fire     10                    N  Defeat                250               2500             14
95    fire     95                    Y Victory                250               2000              4
96    fire     30                    N  Defeat                250               2000             10
97    fire     30                    N  Defeat                500               4000             13
98    fire     20                    N  Defeat                500               4000             13
99    fire     70                    Y  Defeat                500               5000              3
100   fire     90                    Y Victory                500               5000              6
101   fire     30                    N  Defeat                500               5000             12
102   fire     40                    N  Defeat               1000               5000              3
103   fire     45                    N  Defeat               1000               5000              1
104   fire     15                    N  Defeat               1000              10000             12
105   fire     90                    Y Victory               1000              10000              3
106   fire     15                    N  Defeat               1000              10000             10
107   fire     20                    N  Defeat               1500              25000              2
108   fire     25                    N  Defeat               1500              10000              4
109   fire     85                    Y Victory               1500              10000             10
110   fire     85                    Y Victory               2000               7500              9
111   fire     30                    N  Defeat               2000               7500              7
112   fire     40                    N  Defeat               2000              25000              2
113   fire     40                    N  Defeat               2500              20000              3
114   fire     30                    N  Defeat               2500              18000              7
115   fire     80                    Y Victory               2500              12000              4
116   fire     15                    N  Defeat               5000              10000             11
117   fire     10                    N  Defeat               5000              50000             13
118   fire     80                    Y Victory               7500              25000              4
119   fire     50                    Y  Defeat               7500              50000              9
120   fire     15                    N  Defeat              10000              25000             10
```

What just happened?

As we have in previous chapters, we imported our dataset and then created a subset containing our fire attack data. However, this time we used a slightly different function, called `read.table(...)`, to import our external data into R.

read.table(...)

Up to this point, we have always used the `read.csv()` function to import data into R. However, you should know that there are often many ways to accomplish the same objectives in R. For instance, `read.table(...)` is a generic data import function that can handle a variety of file types. While it accepts several arguments, the following three are required to properly import a CSV file, like the one containing our battle history data:

- `file`: the name of the file to be imported, along with its extension, in quotes
- `header`: whether or not the file contains column headings; `TRUE` for yes, `FALSE` (default) for no
- `sep`: the character used to separate values in the file, in quotes

Using these arguments, we were able to import the data in our `battleHistory.csv` into R. Since our file contained headings, we used a value of `TRUE` for the `header` argument and because it is a comma-separated values file, we used `","` for our `sep` argument:

```
> battleHistory <- read.table("battleHistory.csv", TRUE, ",")
```

This is just one example of how a different technique can be used to achieve a similar outcome in R. We will continue to explore new methods in our upcoming activities.

Pop quiz

1. Suppose you wanted to import the following dataset, named *newData* into R. Which of the following `read.table(...)` functions would be best to use?

```
4,5
5,9
6,12
```

a. `read.table("newData", FALSE, ",")`

b. `read.table("newData", TRUE, ",")`

c. `read.table("newData.csv", FALSE, ",")`

d. `read.table("newData.csv", TRUE, ",")`

Time for action – data exploration

To begin our analysis, we will examine the summary statistics and correlations of our data. These will give us an overview of the data and inform our subsequent analyses:

1. Generate a summary of the fire attack subset using `summary(object)`:

```
> #generate a summary of the fire subset
> summaryFire <- summary(subsetFire)
> #display the summary
> summaryFire
```

```
> summaryFire
       Method       Rating      SuccessfullyExecuted      Result    ShuSoldiersEngaged WeiSoldiersEngaged DurationInDays
 ambush    : 0   Min.   : 10.0   N:20                 Defeat :22   Min.   :  100      Min.   : 1500      Min.   : 1.000
 fire      :30   1st Qu.: 25.0   Y:10                 Victory: 8   1st Qu.:  500      1st Qu.: 4250      1st Qu.: 3.000
 headToHead: 0   Median : 55.0                                     Median : 1000      Median : 8750      Median : 6.500
 surround  : 0   Mean   : 52.0                                     Mean   : 2052      Mean   :12333      Mean   : 6.867
                 3rd Qu.: 77.5                                     3rd Qu.: 2375      3rd Qu.:16500      3rd Qu.:10.000
                 Max.   :100.0                                     Max.   :10000      Max.   :50000      Max.   :14.000
```

Before calculating correlations, we will have to convert our nonnumeric data from the `Method`, `SuccessfullyExecuted`, and `Result` columns into numeric form.

For a discussion on converting nonnumeric data, refer to the *Quantifying Categorical Variables* section of *Chapter 4*.

2. Recode the `Method` column using `as.numeric(data)`:

```
> #represent categorical data numerically using
as.numeric(data)
> #recode the Method column into Fire = 1
> numericMethodFire <- as.numeric(subsetFire$Method) - 1
```

3. Recode the `SuccessfullyExecuted` column using `as.numeric(data)`:

```
> #recode the SuccessfullyExecuted column into N = 0 and Y = 1
> numericExecutionFire <-
as.numeric(subsetFire$SuccessfullyExecuted) - 1
```

4. Recode the `Result` column using `as.numeric(data)`:

```
> #recode the Result column into Defeat = 0 and Victory = 1
> numericResultFire <- as.numeric(subsetFire$Result) - 1
```

With the `Method`, `SuccessfullyExecuted`, and `Result` columns coded into numeric form, let us now add them back into our fire dataset.

5. Save the data in our recoded variables back into the original dataset:

```
> #save the data in the numeric Method, SuccessfullyExecuted,
and Result columns back into the fire attack dataset
> subsetFire$Method <- numericMethodFire
> subsetFire$SuccessfullyExecuted <- numericExecutionFire
> subsetFire$Result <- numericResultFire
```

6. Display the numeric version of the fire attack subset. Notice that all of the columns now contain numeric data; it will look like the following:

```
> subsetFire
    Method Rating SuccessfullyExecuted Result ShuSoldiersEngaged WeiSoldiersEngaged DurationInDays
91       1     30                    0      0                100               1500              4
92       1    100                    1      1                100               1500              1
93       1     25                    0      0                100               2500              2
94       1     10                    0      0                250               2500             14
95       1     95                    1      1                250               2000              4
96       1     30                    0      0                250               2000             10
97       1     30                    0      0                500               4000             13
98       1     20                    0      0                500               4000             13
99       1     70                    1      0                500               5000              3
100      1     90                    1      1                500               5000              6
101      1     30                    0      0                500               5000             12
102      1     40                    0      0               1000               5000              3
103      1     45                    0      0               1000               5000              1
104      1     15                    0      0               1000              10000             12
105      1     90                    1      1               1000              10000              3
106      1     15                    0      0               1000              10000             10
107      1     20                    0      0               1500              25000              2
108      1     25                    0      0               1500              10000              4
109      1     85                    1      1               1500              10000             10
110      1     85                    1      1               2000               7500              9
111      1     30                    0      0               2000               7500              7
112      1     40                    0      0               2000              25000              2
113      1     40                    0      0               2500              20000              3
114      1     30                    0      0               2500              18000              7
115      1     80                    1      1               2500              12000              4
116      1     15                    0      0               5000              10000             11
117      1     10                    0      0               5000              50000             13
118      1     80                    1      1               7500              25000              4
119      1     50                    1      0               7500              50000              9
120      1     15                    0      0              10000              25000             10
```

7. Having replaced our original text values in the `SuccessfullyExecuted` and `Result` columns with numeric data, we can now calculate all of the correlations in the dataset using the `cor(data)` function:

```
> #use cor(data) to calculate all of the correlations in the
fire attack dataset
> cor(subsetFire)
```

```
> cor(subsetFire)
                      Method     Rating SuccessfullyExecuted     Result ShuSoldiersEngaged WeiSoldiersEngaged DurationInDays
Method                     1         NA                   NA         NA                 NA                 NA             NA
Rating                    NA  1.0000000           0.91945796  0.9006976        -0.11452122        -0.17682082     -0.4597561
SuccessfullyExecuted      NA  0.9194580           1.00000000  0.8528029         0.08174874         0.02636586     -0.2652026
Result                    NA  0.9006976           0.85280287  1.0000000        -0.03270480        -0.15458357     -0.2514287
ShuSoldiersEngaged        NA -0.1145212           0.08174874 -0.0327048         1.00000000         0.73850610      0.1629878
WeiSoldiersEngaged        NA -0.1768208           0.02636586 -0.1545836         0.73850610         1.00000000      0.1070097
DurationInDays            NA -0.4597561          -0.26520256 -0.2514287         0.16298783         0.10700973      1.0000000
Warning message:
In cor(subsetFire) : the standard deviation is zero
```

Note that the error message and `NA` values in our correlation output result from the fact that our **Method** column contains only a single value. This is irrelevant to our analysis and can be ignored.

What just happened?

Initially, we calculated summary statistics for our fire attack dataset using the `summary(object)` function. From this information, we can derive the following useful insights about our past battles:

- The rating of the Shu army's performance in fire attacks has ranged from 10 to 100, with a mean of 45
- Fire attack plans have been successfully executed 10 out of 30 times (33%)
- Fire attacks have resulted in victory 8 out of 30 times (27%)
- Successfully executed fire attacks have resulted in victory 8 out of 10 times (80%), while unsuccessful attacks have never resulted in victory
- The number of Shu soldiers engaged in fire attacks has ranged from 100 to 10,000 with a mean of 2,052
- The number of Wei soldiers engaged in fire attacks has ranged from 1,500 to 50,000 with a mean of 12,333
- The duration of fire attacks has ranged from 1 to 14 days with a mean of 7

Next, we recoded the text values in our dataset's `Method`, `SuccessfullyExecuted`, and `Result` columns into numeric form. After adding the data from these variables back into our our original dataset, we were able to calculate all of its correlations. This allowed us to learn even more about our past battle data:

- The performance rating of a fire attack has been highly correlated with successful execution of the battle plans (0.92) and the battle's result (0.90), but not strongly correlated with the other variables.

- The execution of a fire attack has been moderately negatively correlated with the duration of the attack, such that a longer attack leads to a lesser chance of success (-0.46).
- The numbers of Shu and Wei soldiers engaged are highly correlated with each other (0.74), but not strongly correlated with the other variables.

The insights gleaned from our summary statistics and correlations put us in a prime position to begin developing our regression model.

Pop quiz

1. Which of the following is a benefit of adding a text variable back into its original dataset after it has been recoded into numeric form?
   a. Calculation functions can be executed on the recoded variable.
   b. Calculation functions can be executed on the other variables in the dataset.
   c. Calculation functions can be executed on the entire dataset.
   d. There is no benefit.

Time for action – model development

Let us continue to the most extensive phase of our data analysis, which consists of developing the optimal regression model for our situation. Ultimately, we want to predict the performance rating of the Shu army under potential fire attack strategies. From our previous exploration of the data, we have reason to believe that successful execution greatly influences the outcome of battle. We can also infer that the duration of a battle has some impact on its outcome. At the same time, it appears that the number of soldiers engaged in battle does not have a large impact on the result. However, since the numbers of Shu and Wei soldiers themselves are highly correlated, there is a potential interaction effect between the two that is worth investigating. We will start by using our insights to create a set of potentially useful models:

1. Use the `glm(formula, data)` function to create a series of potential linear models that predict the `Rating` of battle (dependent variable) using one or more of the independent variables in our dataset. Then, use the `summary(object)` command to assess the statistical significance of each model:

```
> #create a linear regression model using the
glm(formula, data) function
> #predict the rating of battle using execution
```

```
> lmFireRating_Execution <- glm(Rating ~ SuccessfullyExecuted,
data = subsetFire)
> #generate a summary of the model
> lmFireRating_Execution_Summary <-
summary(lmFireRating_Execution)
> #display the model summary
> lmFireRating_Execution_Summary
> #keep execution in the model as an independent variable
```

```
> lmFireRating_Execution_Summary

Call:
glm(formula = Rating ~ SuccessfullyExecuted, data = subsetFire)

Deviance Residuals:
    Min       1Q   Median       3Q      Max
-32.500   -9.500    2.500    6.687   19.250

Coefficients:
                     Estimate Std. Error t value Pr(>|t|)
(Intercept)            25.750      2.648   9.725 1.78e-10 ***
SuccessfullyExecuted   56.750      4.586  12.374 7.20e-13 ***
---
Signif. codes:  0 '***' 0.001 '**' 0.01 '*' 0.05 '.' 0.1 ' ' 1

(Dispersion parameter for gaussian family taken to be 140.2232)

    Null deviance: 25396.7  on 29  degrees of freedom
Residual deviance:  3926.2  on 28  degrees of freedom
AIC: 237.36

Number of Fisher Scoring iterations: 2
```

Our first model used only the successful (or unsuccessful) execution of battle plans to predict the performance of the Shu army in a fire attack. Our summary tells us that execution is an important factor to include in the model.

[For a review of regression model interpretation, refer to the Regression section of *Chapter 5*.]

2. Now, let us examine the impact that the duration of battle has on our model:

```
> #predict the rating of battle using execution and duration
> lmFireRating_ExecutionDuration <-
glm(Rating ~ SuccessfullyExecuted + DurationInDays,
data = subsetFire)
> #generate a summary of the model
> lmFireRating_ExecutionDuration_Summary <-
summary(lmFireRating_ExecutionDuration)
```

```
> #display the model summary
> lmFireRating_ExecutionDuration_Summary
>#keep duration in the model as independent variable
```

```
> lmFireRating_ExecutionDuration_Summary

Call:
glm(formula = Rating ~ SuccessfullyExecuted + DurationInDays,
    data = subsetFire)

Deviance Residuals:
    Min       1Q   Median       3Q      Max
-26.515   -6.360    3.052    8.377   12.905

Coefficients:
                     Estimate Std. Error t value Pr(>|t|)
(Intercept)           38.1253     4.0899   9.322 6.27e-10 ***
SuccessfullyExecuted  52.9484     3.9815  13.299 2.28e-13 ***
DurationInDays        -1.6177     0.4493  -3.600  0.00126 **
---
Signif. codes:  0 '***' 0.001 '**' 0.01 '*' 0.05 '.' 0.1 ' ' 1

(Dispersion parameter for gaussian family taken to be 98.24895)

    Null deviance: 25396.7  on 29  degrees of freedom
Residual deviance:  2652.7  on 27  degrees of freedom
AIC: 227.6

Number of Fisher Scoring iterations: 2
```

This model added the duration of battle to execution as a predictor of the Shu army's rating. Here, we found that duration is also an important predictor that should be included in the model.

3. Next, we will inspect the prospects of including the number of Shu and Wei soldiers as predictors in our model:

```
> #predict the rating of battle using execution, duration,
and the number of Shu and Wei soldiers engaged
> lmFireRating_ExecutionDurationSoldiers <-
glm(Rating ~ SuccessfullyExecuted + DurationInDays +
ShuSoldiersEngaged + WeiSoldiersEngaged, data = subsetFire)
> #generate a summary of the model
> lmFireRating_ExecutionDurationSoldiers_Summary <-
summary(lmFireRating_ExecutionDurationSoldiers)
> #display the model summary
> lmFireRating_ExecutionDurationSoldiers_Summary
> #drop the number of Shu and Wei soldiers from model
as independent variables
```

```
> lmFireRating_ExecutionDurationSoldiers_Summary

Call:
glm(formula = Rating ~ SuccessfullyExecuted + DurationInDays +
    ShuSoldiersEngaged + WeiSoldiersEngaged, data = subsetFire)

Deviance Residuals:
    Min       1Q   Median       3Q      Max
-19.404   -5.672    2.546    6.168   10.559

Coefficients:
                      Estimate Std. Error t value Pr(>|t|)
(Intercept)          41.7874992  3.7429832  11.164 3.32e-11 ***
SuccessfullyExecuted 53.8544374  3.4852025  15.452 2.68e-14 ***
DurationInDays       -1.4309681  0.3968799  -3.606  0.00135 **
ShuSoldiersEngaged   -0.0005492  0.0009647  -0.569  0.57426
WeiSoldiersEngaged   -0.0003340  0.0001861  -1.795  0.08483 .
---
Signif. codes:  0 '***' 0.001 '**' 0.01 '*' 0.05 '.' 0.1 ' ' 1

(Dispersion parameter for gaussian family taken to be 73.72798)

    Null deviance: 25396.7  on 29  degrees of freedom
Residual deviance:  1843.2  on 25  degrees of freedom
AIC: 220.68

Number of Fisher Scoring iterations: 2
```

This time, we added the number of Shu and Wei soldiers into our model, but determined that they were not significant enough predictors of the Shu army's performance. Therefore, we elected to exclude them from our model.

4. Lastly, let us investigate the potential interaction effect between the number of Shu and Wei soldiers:

```
> #investigate a potential interaction effect between the
number of Shu and Wei soldiers
> #center each variable by subtracting its mean from each
of its values
> centeredShuSoldiersFire <- subsetFire$ShuSoldiersEngaged
- mean(subsetFire$ShuSoldiersEngaged)
> centeredWeiSoldiersFire <- subsetFire$WeiSoldiersEngaged
- mean(subsetFire$WeiSoldiersEngaged)
> #multiply the two centered variables to create the
interaction variable
> interactionSoldiersFire <- centeredShuSoldiersFire
* centeredWeiSoldiersFire
> #predict the rating of battle using execution, duration,
and the interaction between the number of Shu and Wei
soldiers engaged
> lmFireRating_ExecutionDurationShuWeiInteraction <-
glm(Rating ~ SuccessfullyExecuted + DurationInDays +
interactionSoldiersFire, data = subsetFire)
> #generate a summary of the model
```

```
lmFireRating_ExecutionDurationShuWeiInteraction_Summary
<- summary(lmFireRating_ExecutionDurationShuWeiInteraction)
> #display the model summary
> lmFireRating_ExecutionDurationShuWeiInteraction_Summary
> #keep the interaction between the number of Shu and Wei
soldiers engaged in the model as an independent variable
```

```
> lmFireRating_ExecutionDurationShuWeiInteraction_Summary

Call:
glm(formula = Rating ~ SuccessfullyExecuted + DurationInDays +
    interactionSoldiersFire, data = subsetFire)

Deviance Residuals:
    Min        1Q    Median        3Q       Max
-18.233    -7.248    1.466     6.452    10.535

Coefficients:
                          Estimate Std. Error t value Pr(>|t|)
(Intercept)              3.737e+01  3.467e+00  10.780 4.34e-11 ***
SuccessfullyExecuted     5.602e+01  3.486e+00  16.071 5.08e-15 ***
DurationInDays          -1.237e+00  3.960e-01  -3.125  0.00434 **
interactionSoldiersFire -1.273e-07  3.717e-08  -3.424  0.00206 **
---
Signif. codes:  0 '***' 0.001 '**' 0.01 '*' 0.05 '.' 0.1 ' ' 1

(Dispersion parameter for gaussian family taken to be 70.319)

    Null deviance: 25396.7  on 29  degrees of freedom
Residual deviance:  1828.3  on 26  degrees of freedom
AIC: 218.43

Number of Fisher Scoring iterations: 2
```

We can see that the interaction effect between the number of Shu and Wei soldiers does have a meaningful impact on our model and should be included as an independent variable.

Note that some statisticians may argue that it is inappropriate to include an interaction variable between the Shu and Wei soldiers in this model, without also including the number of Shu and Wei soldiers alone as variables in the model. In this fictitious example, there is no practically significant difference between these two options, and therefore, the interaction term has been included alone for the sake of simplicity and clarity. However, were you to incorporate interaction effects into your own regression models, you are advised to thoroughly investigate the implications of including or excluding certain variables.

We have identified four potential models. To determine which of these is most appropriate for predicting the outcome of our fire attack, we will use an approach known as **Akaike Information Criterion**, or **AIC**:

```
> #use the AIC(object, ...) function to compare the models
and choose the most appropriate one
> #when comparing via AIC, the lowest value indicates the
best statistical model
> AIC(lmFireRating_Execution, lmFireRating_ExecutionDuration,
lmFireRating_ExecutionDurationSoldiers,
lmFireRating_ExecutionDurationShuWeiInteraction)
> #according to AIC, our model that includes execution, duration, and
the interaction effect is best
```

```
> AIC(lmFireRating_Execution, lmFireRating_ExecutionDuration,
lmFireRating_ExecutionDurationSoldiers, lmFireRating_ExecutionDurationShuWeiInteraction)
                                                df      AIC
lmFireRating_Execution                           3 237.3636
lmFireRating_ExecutionDuration                   4 227.6006
lmFireRating_ExecutionDurationSoldiers           6 220.6781
lmFireRating_ExecutionDurationShuWeiInteraction  5 218.4345
```

The AIC procedure revealed that our model containing execution, duration, and the interaction between the number of Shu and Wei soldiers is the best choice for predicting the performance of the Shu army.

What just happened?

We just completed the process of developing potential regression models and comparing them in order to choose the best one for our analysis. Through this process, we determined that the successful execution, duration, and the interaction between the number of Shu and Wei soldiers engaged were statistically significant independent variables, whereas the number of Shu and Wei soldiers alone were not. By using an AIC test, we were able to determine that the model containing all three statistically significant variables was best for predicting the Shu army's performance in fire attacks. Therefore, our final regression equation is as follows:

```
Rating = 37 + 56 * execution - 1.24 * duration - 0.00000013 *
soldiers interaction
```

For a more detailed discussion of model development, refer to the *Regression* section of *Chapter 5*.

glm(...)

Each of our models in this chapter were created using the `glm(formula, data)` function. For our purposes, this function is identical in structure and very similar in effect to the `lm(formula, data)` function that we are already familiar with from *Chapter 5*. We used `glm(formula, data)` here to demonstrate an alternative R function for creating regression models. In your own work, the appropriate function will be determined by the requirements of your analysis.

You may also have noticed that our `lm(formula, data)` functions listed only the variable names in the `formula` argument. This is a short-hand method for referring to our dataset's column names, as demonstrated by the following code:

```
lmFireRating_ExecutionDuration <- glm(Rating ~
SuccessfullyExecuted + DurationInDays, data = subsetFire)
```

Notice that the `subsetFire$` prefix is absent from each variable name and that the `data` argument has been defined as `subsetFire`. When the `data` argument is used, and the independent variables in the `formula` argument are unique, the `dataset$` prefix may be omitted. This technique has the effect of keeping our code more readable, without changing the results of our calculations.

AIC(object, ...)

AIC can be used to compare regression models. It yields a series of AIC values, which indicate how well our models fit our data. AIC is used to compare multiple models relative to each other, whereby the model with the lowest AIC value best represents our data.

Similar in structure to the `anova(object, ...)` function, the `AIC(object, ...)` function accepts a series of objects (regression models in our case) as input. For example, in `AIC(A, B, C)` we are telling R to compare three objects (`A`, `B`, and `C`) using `AIC`. Thus, our AIC function compared the four regression models that we created:

```
> AIC(lmFireRating_Execution, lmFireRating_ExecutionDuration,
lmFireRating_ExecutionDurationSoldiers,
lmFireRating_ExecutionDurationShuWeiInteraction)
```

As output, `AIC(object, ...)` returned a series of AIC values used to compare our models.

Recall that we compared our regression models in *Chapter 5* using `anova(object, ...)`. To demonstrate an alternative R function that can be used to compare models, we used `AIC(object, ...)` in this activity. The `glm(...)` function coordinates well with `AIC(object, ...)`, hence our decision to use them together in this example. Again, the appropriate techniques to use in your future analyses should be determined by the specific conditions surrounding your work.

Pop quiz

1. When can the `dataset$` prefix be omitted from the variables in the `formula` argument of `lm(formula, data)` and `glm(formula, data)`?
   a. When the data argument is defined.
   b. When the data argument is defined and all of the variables come from different datasets.
   c. When the data argument is defined and all of the variables have unique names.
   d. When the data argument is defined, all of the variables come from different datasets, and all of the variables have unique names.
2. Which of the following is **not** true of the `anova(object, ...)` and `AIC(object, ...)` functions?
   a. Both can be used to compare regression models.
   b. Both receive the same arguments.
   c. Both represent different statistical methods.
   d. Both yield identical mathematical results.

Time for action – model deployment

Having selected the optimal model for predicting the outcome of our fire attack strategy, it is time to put that model to practical use. We can use it to predict the outcomes of various fire attack strategies and to identify one or more strategies that are likely to lead to victory. Subsequently, we need to ensure that our winning strategies are logistically sound and viable. Once we strike a balance between our designed strategy and our practical constraints, we will arrive at the best course of action for the Shu forces.

Recall from *Chapter 6* that we set a rating value of 80 as our minimum threshold. As such, we will only consider a strategy adequate if it yields a rating of 80 or higher when all variables have been entered into our model.

In the case of our fire attack regression model, we know that to achieve our desired rating value, we must assume successful execution. We also know the number of Wei soldiers housed at the target city. Consequently, our major constraints are the number of Shu soldiers that we choose to engage in battle and the duration of the attack. We will assume a moderate attack duration.

Subsequently, we can rearrange our regression equation to solve for the number of Shu soldiers engaged and then represent it as a custom function in R:

1. Use the `coef(object)` function to isolate the independent variables in our regression model:

```
> #use the coef(object) function to extract the coefficients
from a regression model
> #this will make it easier to rearrange our equation by
allowing us to focus only on these values
> coef(lmFireRating_ExecutionDurationShuWeiInteraction)
```

```
> coef(lmFireRating_ExecutionDurationShuWeiInteraction)
          (Intercept)    SuccessfullyExecuted          DurationInDays interactionSoldiersFire
         3.737354e+01            5.601947e+01           -1.237476e+00           -1.272603e-07
```

2. Rewrite the fire attack regression equation to solve for the number of Shu soldiers engaged in battle:

```
> #rewrite the regression equation to solve for the number of
Shu soldiers engaged in battle
> #original equation: rating = 37 + 56 * execution - 1.24 *
duration - 0.00000013 * soldiers interaction
> #rearranged equation: Shu soldiers = (rating - 37 + 56 *
execution + 1.24 * duration) / (0.00000013 * -
Wei soldiers engaged)
```

3. Use the `function()` command to create a custom R function to solve for the number of Shu soldiers engaged in battle, given the desired `rating`, `execution`, `duration`, and number of `WeiSoldiers`:

```
> #use function() to create a custom function in R
> #the function() command follows this basic format:
+   function(argument1, argument2,... argumenti) { equation }
> #custom function that solves for the maximum number of Shu
soldiers that can be deployed, given the desired rating,
execution, duration, and number of Wei soldiers
> functionFireShuSoldiers <- function(rating, execution,
duration, WeiSoldiers) {
+      (rating - 37 - 56 * execution +
+      1.24 * duration) /
+      (0.00000013 * - WeiSoldiers)
+ }
```

4. Use the custom function to solve for the number of Shu soldiers that can be deployed, given a `rating` of 80, `duration` of 7, `success` of 1.0, and 10,000 `WeiSoldiers`:

```
> #solve for the number of Shu soldiers that can be deployed
given a result of 80, duration of 7, success of 1.0, and
15,000 WeiSoldiers
> functionFireShuSoldiers(80, 1.0, 7, 10000)
[1] 3323.077
```

Our regression model suggests that to achieve a rating of 80, our minimum threshold, we should deploy 3,323 Shu soldiers. However, from looking at the data in our fire attack subset, a force between 2,500 and 5,000 soldiers has not been previously used to launch a fire attack. Further, four past successful fire attacks on 7,500 to 12,000 Wei soldiers have deployed only 1,000 to 2,500 Shu soldiers. What would happen to our predicted rating value if we were to deploy 2,500 Shu soldiers instead of 3,323?

1. Create a custom function to solve for the rating of battle when `execution`, `duration`, and number of `ShuSoldiers` and `WeiSoldiers` are known:

```
> #custom function that solves for rating of battle, given the
execution, duration, number of Shu soldiers, and number of Wei
soldiers
> functionFireRating <- function(execution, duration,
ShuSoldiers, WeiSoldiers) {
+    37 + 56 * execution -
+    1.24 * duration -
+    0.00000013 * (ShuSoldiers * WeiSoldiers)
+ }
```

2. Use the custom function to solve for the rating of battle, given successful execution, a 7-day duration, 2,500 Shu soldiers, and 10,000 Wei soldiers:

```
> What would happen to our rating value if we were to deploy
2,500 Shu soldiers instead of 3,323?
> functionFireRating(1.0, 7, 2500, 10000)
[1] 81.07
> #Is the 1.07 increase in our predicted chances for victory
worth the practical benefits derived from deploying 2,500
soldiers?
```

By using 2,500 soldiers, our rating value increased to 81, which is slightly above our threshold of confidence for victory. Here, we have encountered a classic dilemma for the data analyst. On one hand, our data model tells us that it is safe to use 3,323 soldiers. On the other, our knowledge of war strategy and past outcomes tells us that a number between 1,000 and 2,500 would be sufficient. Essentially, we have to identify the practical benefits or detriments from deploying a certain number of soldiers. In this case, we are inclined to think that it is beneficial to deploy fewer than 3,323, but more than 1,000. The exact number is a matter of debate and uncertainty that deserves serious consideration. It is always the strategist's challenge to weigh both the practical and statistical benefits of potential decisions. On that note, let us consider the logistics of our proposed fire attack. Our plan is to deploy 2,500 Shu soldiers over a period of 7 days to attack 10,000 Wei soldiers who are stationed 225 miles away.

1. Create a custom function that calculates the gold cost of our fire attack strategy:

```
> #custom function that calculates the gold cost of our
strategy, given the number of Shu soldiers deployed, the
distance of the target city, and the proposed duration of
battle.
> functionGoldCost <- function(ShuSoldiers, distance, duration)
+ {
+    ShuSoldiers * (distance / 100 + 2 * (duration / 30))
+ }
```

2. Use the custom function to calculate the gold cost of our fire attack strategy:

```
> #gold cost of fire attack that deploys 2,500 Shu soldiers
a distance of 225 miles for a period of 7 days
> functionGoldCost(2500, 225, 7)
[1] 6791.667
```

3. Calculate the number of provisions needed for our fire attack strategy:

```
> #provisions required by our fire attack strategy
> #consumption per 30 days is equal to the number of soldiers
deployed
> 2500 * (7/30)
[1] 583.3333
```

4. Determine whether the fire attack strategy is viable given our resource limitations:

```
> #our gold cost of 6,792 is well below our allotment of 1,000,000
> #our required provisions of 583 are well below our allotment of
1,000,000
> #our 2,500 soldiers account for only 1.25% of our total army
personnel
> #yes, the fire attack strategy is viable given our resource
constraints
```

What just happened?

We successfully used our optimal regression model to refine our battle strategy and test its viability in light of our practical resource constraints. Custom functions were used to calculate the number of soldiers necessary to yield our desired outcome, the performance rating given the parameters of our plan, and the overall gold cost of our strategy. In determining the number of soldiers to engage in our fire attack, we encountered a common occurrence whereby our data models conflicted with our practical understanding of the world. Subsequently, we had to use our expertise as data analysts to balance the consequences between the two and arrive at a sound conclusion. We then assessed the overall viability of our strategy and determined it to be sufficient in consideration of our resource allotments.

For a more detailed discussion of the techniques used in this segment, refer to the *Logistical Considerations* section of *Chapter 6*.

coef(object)

Prior to rewriting our regression equation and converting it into a custom function, we executed the `coef(object)` command on our model. The `coef(object)` function, when executed on a regression model, has the effect of extracting and displaying its independent variables (or **coefficients**). By isolating these components, we were able to easily visualize our model's equation:

```
> coef(lmFireRating_ExecutionDurationShuWeiInteraction)
```

```
> coef(lmFireRating_ExecutionDurationShuWeiInteraction)
    (Intercept)    SuccessfullyExecuted          DurationInDays interactionSoldiersFire
   3.737354e+01            5.601947e+01           -1.237476e+00           -1.272603e-07
```

In contrast, the `summary(object)` function contains much more information than we need for this purpose, thus making it potentially confusing and difficult to locate our variables. This can be seen in the following:

```
> lmFireRating_ExecutionDurationShuWeiInteraction_Summary
```

```
> lmFireRating_ExecutionDurationShuWeiInteraction_Summary

Call:
glm(formula = Rating ~ SuccessfullyExecuted + DurationInDays +
    interactionSoldiersFire, data = subsetFire)

Deviance Residuals:
    Min       1Q   Median       3Q      Max
-18.233   -7.248    1.466    6.452   10.535

Coefficients:
                         Estimate Std. Error t value Pr(>|t|)
(Intercept)             3.737e+01  3.467e+00  10.780 4.34e-11 ***
SuccessfullyExecuted    5.602e+01  3.486e+00  16.071 5.08e-15 ***
DurationInDays         -1.237e+00  3.960e-01  -3.125  0.00434 **
interactionSoldiersFire -1.273e-07  3.717e-08  -3.424  0.00206 **
---
Signif. codes:  0 '***' 0.001 '**' 0.01 '*' 0.05 '.' 0.1 ' ' 1

(Dispersion parameter for gaussian family taken to be 70.319)

    Null deviance: 25396.7  on 29  degrees of freedom
Residual deviance:  1828.3  on 26  degrees of freedom
AIC: 218.43

Number of Fisher Scoring iterations: 2
```

Hence, in circumstances where we only care to see the independent variables in our model, the `coef(object)` function can be more effective than `summary(object)`.

Pop quiz

1. Under which of the following circumstances might you use the `coef(object)` function instead of `summary(object)`?

   a. You want to know the practical significance of the model's variables.

   b. You want to know the statistical significance of the model's variables.

   c. You want to know the model's regression equation.

   d. You want to know the formula used to generate the model.

Time for action – last steps

Lastly, we need to save the workspace and console text associated with our fire attack analysis:

1. Use the `save.image(file)` function to save your R workspace to your working directory. The `file` argument should contain a meaningful filename and the `.RData` extension:

   ```
   > #save the R workspace to your working directory
   > save.image("rBeginnersGuide_Ch_07_fireAttackAnalysis.RData")
   ```

2. R will save your workspace file. Browse to the working directory on your hard drive to verify that this file has been created.

3. Manually save your R console log by copying and pasting it into a text file. You may then format the console text to improve its readability.

We have now completed an entire data analysis of the fire attack strategy from beginning to end using R.

The common steps to all R analyses

While retracing the development process behind our fire attack strategy, we encountered a key series of steps that are common to every analysis that you will conduct in R. Regardless of the exact situation or the statistical techniques used, there are certain things that must be done to yield an organized and thorough R analysis. Each of these steps is detailed.

Perhaps it goes without saying that the thing to do before beginning any R analysis is to launch R itself. Nevertheless, it is mentioned here for completeness and transparency.

Step 1: Set your working directory

Once R is launched, the first common step is to set your working directory. This can be done using the `setwd(dir)` function and subsequently verified using the `getwd()` command:

```
> #Step 1: set your working directory

> #set your working directory using setwd(dir)
> #replace the sample location with one that is relevant to you
> setwd("/Users/johnmquick/rBeginnersGuide/")

> #once set, you can verify your new working directory using getwd()
> getwd()
[1] "/Users/johnmquick/rBeginnersGuide/"
```

Comment your work

Note that commented lines, which are prefixed with the pound sign (#), appeared before each of our functions in step one. It is vital that you comment all of the actions that you take within the R console. This allows you to refer back to your work later and also makes your code accessible to others.

This is an opportune time to point out that you can draft your code in other places besides the R console. For example, R has a built in editor that can be opened by going to the **File | New Document/Script** menu or simultaneously pressing the *Command* + *N* or *Ctrl* + *N* keys. Other free editors can also be found online. The advantages of using an editor are that you can easily modify your code and see different types of code in different colors, which helps you to verify that it is properly constructed. Note however, that to execute your code, it must be placed in the R console.

Step 2: Import your data (or load an existing workspace)

After you set the working directory, it is time to pull your data into R. This can be achieved by creating a new variable in tandem with the `read.csv(file)` command:

```
> #Step 2: Import data (or load an existing workspace)

> #read a dataset from a csv file into R using read.csv(file) and save
it into a new variable
> dataset <- read.csv("datafile.csv")
```

Alternatively, if you were continuing a prior data analysis, rather than starting a new one, you would instead load a previously saved workspace using `load.image(file)`. You can then verify the contents of your loaded workspace using the `ls()` command.

```
> #load an existing workspace using load.image(file)
> load.image("existingWorkspace.RData")

> #verify the contents of your workspace using ls()
> ls()
[1] "myVariable 1"
[2] "myVariable 2"
[3] "myVariable 3"
```

Step 3: Explore your data

Regardless of the type or amount of data that you have, summary statistics should be generated to explore your data. Summary statistics provide you with a general overview of your data and can reveal overarching patterns, trends, and tendencies across a dataset. Summary statistics include calculations such as means, standard deviations, and ranges, amongst others:

```
> #Step 3: Explore your data

> #calculate a mean using mean(data)
> mean(myData)
[1] 1000

> #calculate a standard deviation using sd(data)
> sd(myData)
[1] 100

> #calculate a range (minimum and maximum) using range(data)
> range(myData)
> [1] 500 2000
```

Also recall R's `summary(object)` function, which provides summary statistics along with additional vital information. It can be used with almost any object in R and will offer information specifically catered to that object:

```
> #generate a detailed summary for a given object using
summary(object)
> summary(object)
```

Note that there are often other ways to make an initial examination of your data in addition to using summary statistics. When appropriate, graphing your data is an excellent way to gain a visual perspective on what it has to say (data visualization is the primary topic of *Chapter 8* and *Chapter 9* of this book). Furthermore, before conducting an analysis, you will want to ensure that your data are consistent with the assumptions necessitated by your statistical methods. This will prevent you from expending energy on inappropriate techniques and from making invalid conclusions.

Step 4: Conduct your analysis

Here is where your work will differ from project to project. Depending on the type of analysis that you are conducting, you will use a variety of different techniques. For example, in this book we have primarily used regression analysis. Regression is but one of an endless number of potential methods. The correct techniques to use will be determined by the circumstances surrounding your work.

```
> #Step 4: Conduct your analysis
> #The appropriate methods for this step will vary between analyses.
```

Step 5: Save your workspace and console files

At the conclusion of your analysis, you will always want to save your work. To have the option to revisit and manipulate your R objects from session to session, you will need to save your R workspace using the `save.image(file)` command, as follows:

```
> #Step 5: Save your workspace and console files

> #save your R workspace using save.image(file)
> #remember to include the .RData file extension
> save.image("myWorkspace.RData")
```

To save your R console text, which contains the log of every action that you took during a given session, you will need to copy and paste it into a text file. Once copied, the console text can be formatted to improve its readability. For instance, a text file containing the five common steps of every R analysis could take the following form:

```
> #There are five steps that are common to every data analysis
conducted in R

> #Step 1: set your working directory

> #set your working directory using setwd(dir)
> #replace the sample location with one that is relevant to you
> setwd("/Users/johnmquick/rBeginnersGuide/")

> #once set, you can verify your new working directory using getwd()
> getwd()
[1] "/Users/johnmquick/rBeginnersGuide/"

> #Step 2: Import data (or load an existing workspace)

> #read a dataset from a csv file into R using read.csv(file) and save
it into a new variable
> dataset <- read.csv("datafile.csv")
```

```
> #OR

> #load an existing workspace using load.image(file)
> load.image("existingWorkspace.RData")

> #verify the contents of your workspace using ls()
> ls()
[1] "myVariable 1"
[2] "myVariablc 2"
[3] "myVariable 3"

> #Step 3: Explore your data

> #calculate a mean using mean(data)
> mean(myData)
[1] 1000

> #calculate a standard deviation using sd(data)
> sd(myData)
[1] 100

> #calculate a range (minimum and maximum) using range(data)
> range(myData)
> [1] 500 2000

> #generate a detailed summary for a given object using
summary(object)
> summary(object)

> #Step 4: Conduct your analysis
> #The appropriate methods for this step will vary between analyses.

> #Step 5: Save your workspace and console files

> #save your R workspace using save.image(file)
> #remember to include the .RData file extension
> save.image("myWorkspace.RData")

> #save your R console text by copying it and pasting it into a text
file.
```

See the `rBeginnersGuide_CommonSteps.txt` file that is provided with this book.

Pop quiz

1. Which of the following is not a benefit of commenting your code?
   a. It makes your code readable and organized.
   b. It makes your code accessible to others.
   c. It makes it easier for you to return to and recall your past work.
   d. It makes the analysis process faster.

Have a go hero

Conduct a complete end to end analysis using the strategy that you decided upon at the conclusion of *Chapter 6*. Be sure to employ each of the five common steps to all R analyses. Along the way, refer to the *Retracing and Refining a Complete Analysis* section of this chapter, as well as the previous chapters of this book. Once your analysis is complete, you should have the following items:

- A workspace file containing all of the objects used in your analysis
- A commented console text file detailing all of the actions that occurred during your analysis
- A sound, viable battle strategy for the Shu army

Summary

In this chapter, we conducted an entire data analysis in R from beginning to end. While doing so, we ensured that our work was as organized and transparent as possible, thereby making it more accessible to others. Afterwards, we identified the five steps that are common to all well-executed data analyses in R. You then used these steps to conduct, organize, and refine a battle strategy for the Shu army. Having completed this chapter, you should now be able to:

- Organize and clarify your raw R data analyses
- Communicate your raw R data analyses effectively
- Apply the steps common to all well-conducted R analyses

Now that we have a complete, organized, and clear plan for the Shu army, our challenge is to communicate it to others. Next, we will visit the Emperor, who has the power to accept or reject our battle plans. In order to communicate our ideas simply and effectively, we will focus on using graphical techniques in *Chapter 8*.

8
Briefing the Emperor

You revisited and reorganized a complete data analysis in Chapter 7 to prepare your strategy for presentation. The next step towards executing your plans for the Shu forces is to visit the emperor and propose your strategy. The emperor is unconcerned with the minute details of the attack, but rather needs to be convinced that your proposed attack is a sound one that will be beneficial for the Shu kingdom. It is important to convey your plans with clarity, because the emperor has the power accept or reject your strategy.

To provide the emperor with the clear and concise information that he needs, we will focus on R's graphical features. We will convey our strategy through the use of several charts, graphs, and plots. We will also explore our options for customizing these visuals. Through the use of R's rich graphical features, we make the benefits of our combat strategy readily apparent and win the support of the emperor. By the end of this chapter, you will be able to:

- Create six different charts, graphs, and plots in R
- Customize your R visuals using text, colors, axes, and legends
- Save and export your graphics for use outside of R

Charts, graphs, and plots in R

R features several options for creating charts, graphs, and plots. In this chapter, we will explore the generation and customization of these visuals, as well as methods for saving and exporting them for use outside of R. The following visuals will be covered in this chapter:

- Bar graphs
- Scatterplots

- Line charts
- Box plots
- Histograms
- Pie charts

For demonstration purposes, all of our visuals will communicate information about the fire attack strategy that was used in *Chapter 7*. This strategy entailed deploying 2,500 Shu soldiers for 7 days to execute a fire attack 225 miles away on 10,000 Wei soldiers in And. If desired, you are encouraged to substitute your own battle plans into any or all activities for this chapter.

Time for action – creating a bar chart

A **bar chart** or **bar graph** is a common visual that uses rectangles to depict the values of different items. Bar graphs are especially useful when comparing data over time or between diverse groups. Let us create a bar chart in R:

1. Open R and set your working directory:

```
> #set the R working directory
> #replace the sample location with one that is relevant to you
> setwd("/Users/johnmquick/rBeginnersGuide/")
```

2. Load the Chapter 8 workspace. It contains the necessary information for this chapter.

```
> #load the chapter 8 workspace
> load("rBeginnersGuide_Ch_08_ReadersCopy.RData")
```

3. Use the `barplot(...)` function to create a bar chart:

```
> #create a bar chart that compares the mean durations of
the battle methods
> #calculate the mean duration of each battle method
> meanDurationFire <- mean(subsetFire$DurationInDays)
> meanDurationAmbush <- mean(subsetAmbush$DurationInDays)
> meanDurationHeadToHead <-
mean(subsetHeadToHead$DurationInDays)
> meanDurationSurround <- mean(subsetSurround$DurationInDays)
> #use a vector to define the chart's bar values
> barAllMethodsDurationBars <- c(meanDurationFire,
meanDurationAmbush, meanDurationHeadToHead,
meanDurationSurround)
> #use barplot(...) to create and display the bar chart
> barplot(height = barAllMethodsDurationBars)
```

4. Your chart will be displayed in the graphic window, similar to the following:

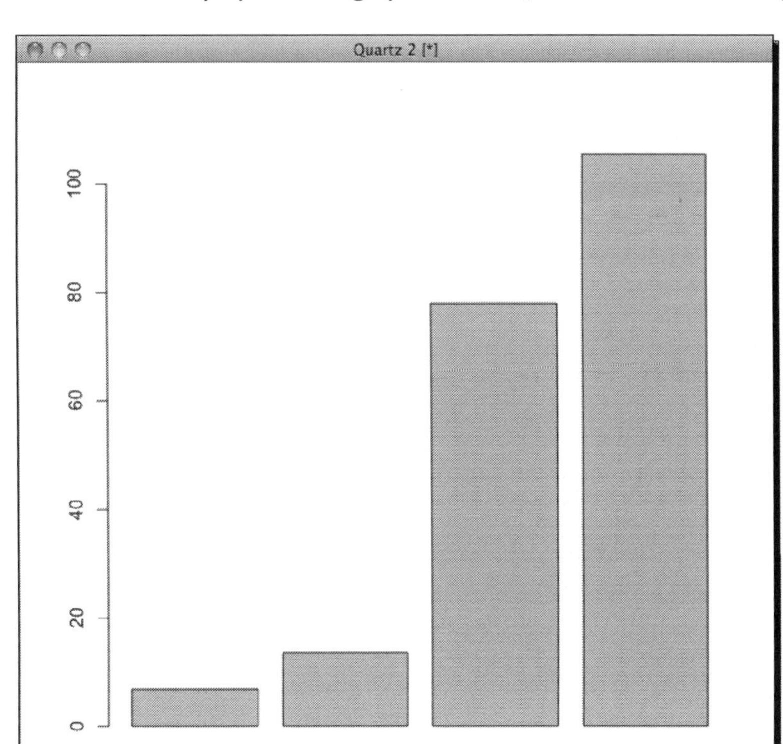

What just happened?

You created your first graphic in R. Let us examine the `barplot(...)` function that we used to generate our bar chart, along with the new R components that we encountered.

barplot(...)

We created a bar chart that compared the mean durations of battles between the different combat methods. As it turns out, there is only one required argument in the `barplot(...)` function. This `height` argument receives a series of values that specify the length of each bar. Therefore, the `barplot(...)` function, at its simplest, takes on the following form:

```
barplot(height = heightValues)
```

Accordingly, our bar chart function reflected this same format:

```
> barplot(height = barAllMethodsDurationBars)
```

Vectors

We stored the heights of our chart's bars in a **vector** variable. In R, a vector is a series of data. R's `c(...)` function can be used to create a vector from one or more data points. For example, the numbers 1, 2, 3, 4, and 5 can be arranged into a vector like so:

```
> #arrange the numbers 1, 2, 3, 4, and 5 into a vector
> numberVector <- c(1, 2, 3, 4, 5)
```

Similarly, text data can also be placed into vector form, so long as the values are contained within quotation marks:

```
> #arrange the letters a, b, c, d, and e into a vector
> textVector <- c("a", "b", "c", "d", "e")
```

Our vector defined the values for our bars:

```
> #use a vector to define the chart's bar values
> barAllMethodsDurationBars <- c(meanDurationFire,
meanDurationAmbush, meanDurationHeadToHead, meanDurationSurround)
```

Many function arguments in R require vector input. Hence, it is very common to use and encounter the `c(...)` function when working in R.

Graphic window

When you executed your `barplot(...)` function in the R console, the **graphic window** opened to display it. The graphic window will have different names across different operating systems, but its purpose and function remain the same. For example, in Mac OS X, the graphic window is named *Quartz*.

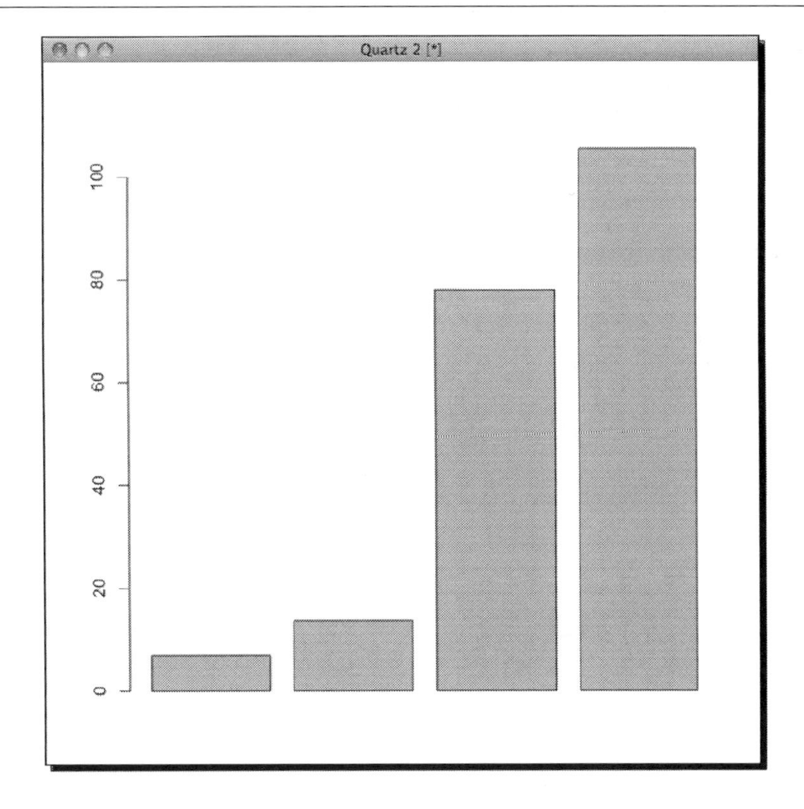

For the remainder of this book, all R graphics will be displayed without the graphics window frame, which will allow us to focus on the visuals themselves.

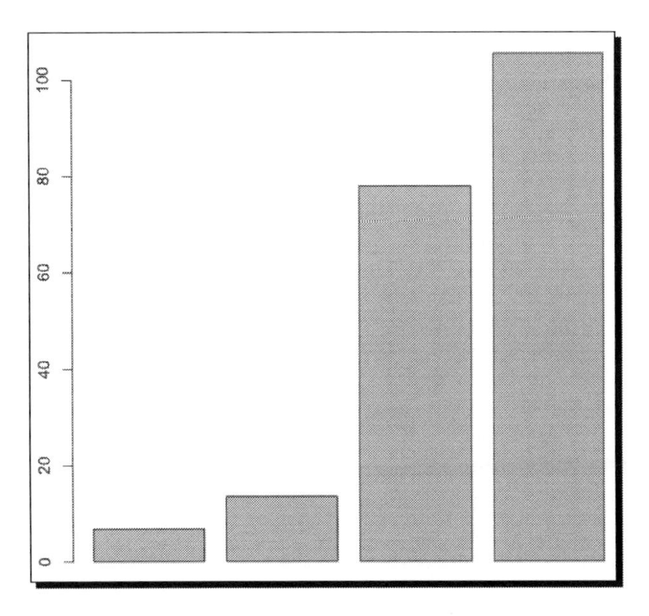

Pop quiz

1. When entering text into a vector using the `c(...)` function, what characters must surround each text value?
   a. quotation marks
   b. parenthesis
   c. asterisks
   d. percent signs
2. What is the purpose of the R graphic window?
   a. to debug graphics functions
   b. to execute graphics functions
   c. to edit graphics
   d. to display graphics

Time for action – customizing graphics

Although the `barplot(...)` function only requires the `height` of each bar to be specified, creating a chart in this manner leaves us with a bland and difficult to decipher visual. In most cases, you will want to customize your R graphics by incorporating additional arguments into your functions. Let us explore how to use graphic customization arguments by expanding our bar chart:

1. Expand your bar chart using graphic customization arguments:

```
> #use additional arguments to customize a graphic
> #define a title for the bar chart
> barAllMethodsDurationLabelMain <-
"Average Duration by Battle Method"
> #define x and y axis labels for the bar chart
> barAllMethodsDurationLabelX <- "Battle Method"
> barAllMethodsDurationLabelY <- "Duration in Days"
> #set the x and y axis scales
> barAllMethodsDurationLimX <- c(0, 5)
> barAllMethodsDurationLimY <- c(0, 120)
> #define rainbow colors for the bars
> barAllMethodsDurationRainbowColors <-
rainbow(length(barAllMethodsDurationBars))
> #incorporate customizations into the graphic function using
the main, xlab, ylab, xlim, ylim, names, and col arguments
> #use barplot(...) to create and display the bar chart
```

```
> barplot(height = barAllMethodsDurationBars,
main = barAllMethodsDurationLabelMain,
xlab = barAllMethodsDurationLabelX,
ylab = barAllMethodsDurationLabelY,
xlim = barAllMethodsDurationLimX,
ylim = barAllMethodsDurationLimY,
col = barAllMethodsDurationRainbowColors)
```

2. Your chart will be displayed in the graphic window, as shown in the following screenshot:

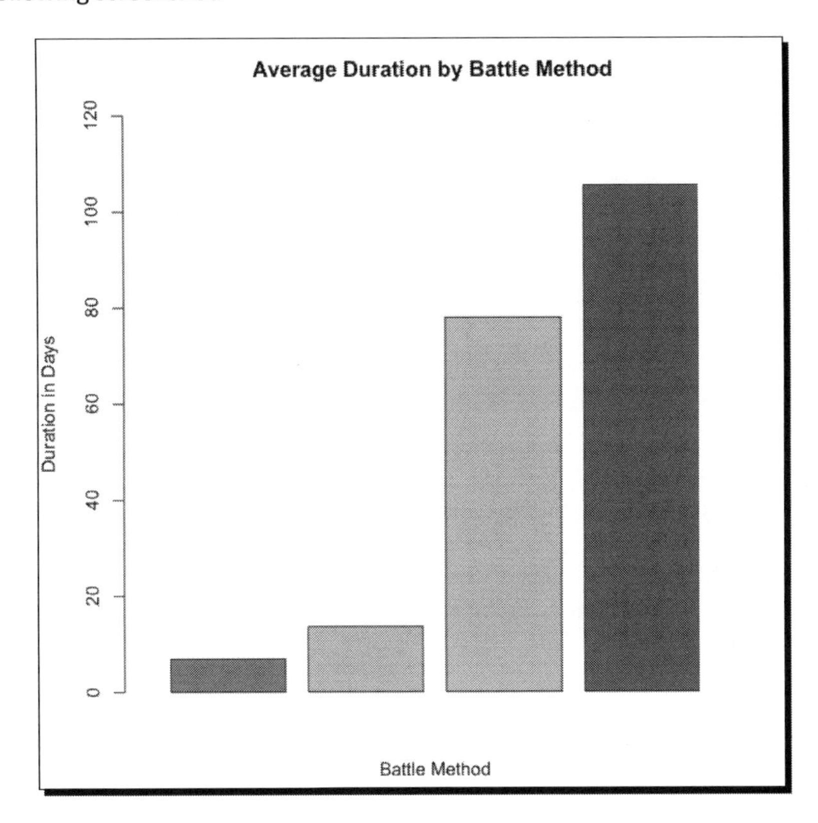

3. Add a legend to the chart, using the following snippet:

```
> #add a legend to the bar chart
> #the x and y arguments position the legend
> #x and y can be defined using words or numerical coordinates
> #the legend argument receives a vector containing the labels
for the legend
> barAllMethodsDurationLegendLabels <- c("Fire", "Ambush",
"Head to Head", "Surround")
```

```
> #the fill argument contains the colors for the legend
> legend(x = 0, y = 120,
legend = barAllMethodsDurationLegendLabels,
fill = barAllMethodsDurationRainbowColors)
```

4. Your legend will be added to the existing chart.

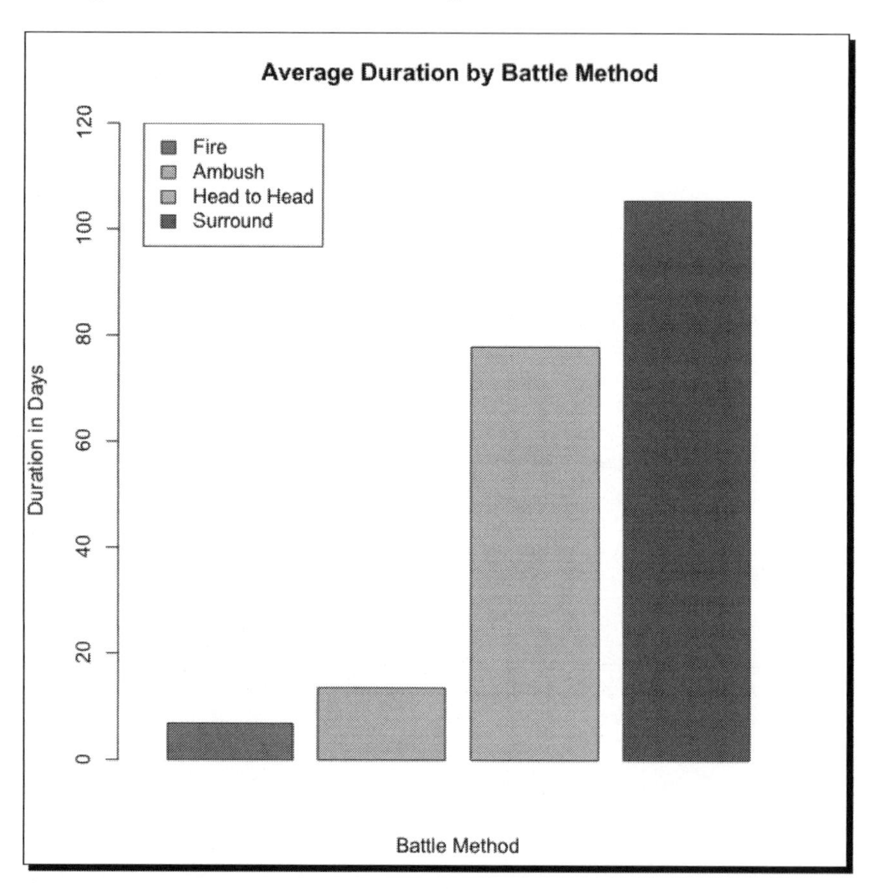

What just happened?

The `barplot(...)` function, as well as the other graphic functions that we will use in this book, accept a variable number of arguments. In fact, R graphics functions have many customizable options and therefore tend to accept several arguments. We expanded our bar chart using a collection of the most common customization arguments, which apply to nearly all R graphics functions.

Graphic customization arguments

We used six arguments to customize our bar chart:

- `main`: a text title for the graphic
- `xlab`: a text label for the x axis
- `ylab`: a text label for the y axis
- `xlim`: a vector containing the lower and upper limits for the x axis
- `ylim`: a vector containing the lower and upper limits for the y axis
- `col`: a vector containing the colors to be used in the graphic

The general format for these arguments is as follows:

```
argument = value
```

When incorporated into a graphics function, these arguments take on the following form:

```
graphicsFunction(..., argument = value)
```

Recognize that these six arguments can be applied to nearly every R graphics function. Each one can be used alone or they can be used in tandem. We will use these arguments throughout the chapter to refine and improve our visuals.

main, xlab, and ylab

The `main`, `xlab`, and `ylab` arguments are all used to add clarifying text to graphics. A primary title for a graphic is defined by `main`, while labels for the x and y axes are specified using `xlab` and `ylab`, respectively.

Our `barplot(...)` function made use of the `main`, `xlab`, and `ylab` arguments. We saved our argument values into variables prior to incorporating them into the `barplot(...)` function. First, we defined our text values as variables.

```
> #define a title for the bar chart
> barAllMethodsDurationLabelMain <-
"Average Duration by Battle Method"
> #define x and y axis labels for the bar chart
> barAllMethodsDurationLabelX <- "Battle Method"
> barAllMethodsDurationLabelY <- "Duration in Days"
```

Then, we used our variables in the final `barplot(...)` function:

```
> barplot(height = barAllMethodsDurationBars,
main = barAllMethodsDurationLabelMain,
xlab = barAllMethodsDurationLabelX,
ylab = barAllMethodsDurationLabelY,
xlim = barAllMethodsDurationLimX,
ylim = barAllMethodsDurationLimY,
col = barAllMethodsDurationRainbowColors)
```

This variable technique has the advantages of rendering our code more decipherable and making it easier for us to return to and reuse our data in future graphics. We will continue to use this method throughout the chapter.

xlim and ylim

The `xlim` and `ylim` arguments receive a vector containing the minimum and maximum values for the x and y axes respectively. Thus, in:

```
xlim = c(50, 250)
```

A graphic's x axis is told to present the data that fall between 50 and 250. The `ylim` argument operates in identical fashion to `xlim`, with the exception that it acts upon the y axis. These arguments are useful for rescaling a graphic's axes to improve its visual presentation. They can also have the effect of emphasizing or deemphasizing certain data ranges.

In our chart, we used `xlim` to set a minimum of 0 and a maximum of 5 for the x axis. This evenly and comfortably spaced our bars within the graphic window. We used `ylim` to set a minimum of 0 and maximum of 120 for the y axis. This ensured that all of our data were represented and that our bars were displayed at a reasonable height.

```
> barplot(height = barAllMethodsDurationBars,
main = barAllMethodsDurationLabelMain,
xlab = barAllMethodsDurationLabelX,
ylab = barAllMethodsDurationLabelY,
xlim = barAllMethodsDurationLimX,
ylim = barAllMethodsDurationLimY,
col = barAllMethodsDurationRainbowColors)
```

Col

R can generate colors in two different forms using Col; they can be rainbow colors which are automatic, or you can specify colors of your choice.

Rainbow colors

R can generate an automatic sequence of colors for a chart with the `rainbow(...)` function. For our purposes, we simply identified the number of colors that we wished to generate for our chart. To obtain the appropriate number of colors, we used the `length(object)` command. This function tells us the number of items contained in a given object. In our case, using `length(object)` on the `barAllMethodsDurationBars` yielded a result of 4, which represents each of our chart's bars:

```
> barAllMethodsDurationSpecificColors <-
rainbow(length(barAllMethodsDurationBars))
```

Consequently, the `rainbow(...)` function generated four colors. These colors were applied to the chart's bars when we included the `barAllMethodsDurationRainbowColors` variable in the `col` argument of our `barplot(...)` function.

```
> barplot(height = barAllMethodsDurationBars,
main = barAllMethodsDurationLabelMain,
xlab = barAllMethodsDurationLabelX,
ylab = barAllMethodsDurationLabelY,
xlim = barAllMethodsDurationLimX,
ylim = barAllMethodsDurationLimY,
col = barAllMethodsDurationRainbowColors)
```

Specific colors

Alternatively, specific colors can be defined using the `col` argument in tandem with a vector list of color names. Common color names such as red, green, blue, and yellow are valid inputs. In this situation, the `col` argument takes on the following form:

```
col = colorVector
```

Where `colorVector` is a variable storing a vector of color values like the following:

```
c("red", "green", "blue", "yellow")
```

You can see a complete list of the colors available in R by executing the `colors()` function.

Had we wanted to use specific colors in our bar chart, we could have employed the following code:

```
> #define specific colors for the bars
> barAllMethodsDurationSpecificColors <- c("red", "green", "blue",
"yellow")
> #use barplot(...) to create and display the bar chart
> barplot(height = barAllMethodsDurationBars,
main = barAllMethodsDurationLabelMain,
xlab = barAllMethodsDurationLabelX,
ylab = barAllMethodsDurationLabelY,
xlim = barAllMethodsDurationLimX,
ylim = barAllMethodsDurationLimY,
col = barAllMethodsDurationSpecificColors)
```

legend(...)

The finishing touch to our bar chart was a legend, or key, that indicated what our bars represented. In R, the `legend(...)` function employs the following arguments:

- `x`: the x position of the chart in numeric terms; alternatively you can set the overall position of the legend using one of the text values `topleft`, `top`, `topright`, `left`, `center`, `right`, `bottomleft`, `bottomcenter`, or `bottomright`
- `y`: the y position of the chart in numeric terms; if text is used for x, omit this argument
- `legend`: a vector containing the labels to be used in the legend
- `fill`: a vector containing the colors to be used in the legend

The basic format for the legend function is as follows:

```
legend(x = xPosition, y = yPosition, legend = labelVector,
fill = colorVector)
```

For instance, the following code:

```
> legend(x = "topleft", legend = c("a", "b"), fill = rainbow(2))
```

This would yield a legend placed at the top-left position with labels for `a` and `b` whose colors were generated by the `rainbow(...)` function. Note that the `x` argument used a text value and `y` was omitted as an alternative to defining the exact numerical position of the legend.

Our function used the x and y coordinates from our chart to position the legend in the upper left-hand corner. When using numbers to define the `x` and `y` arguments, the values will always depend on the limits of the x and y axes. For instance, a position of (0, 120) specified the upper left-hand corner in our chart, but a graphic with a maximum y value of 50 would have an upper left-hand corner position of (0, 50). Our `legend` and `fill` arguments incorporated the same labels and colors that were used to generate our bar chart. Thus, our legend was matched to the information depicted in our chart:

```
> legend(x = 0, y = 120,
legend = barAllMethodsDurationLegendLabels,
fill = barAllMethodsDurationRainbowColors)
```

Notice the peculiar implementation of the `legend(...)` function, which we have not previously encountered. As we will see with other graphics functions, `legend(...)` does not stand alone. To be properly employed, a compatible graphic must already exist for `legend(...)` to act upon. In this situation, `legend(...)` adds a new legend *on top of the visual that is displayed in the graphic window*. However, if no graphic is currently displayed when the `legend(...)` function is executed, an error message is returned. This is demonstrated in the following code:

```
> #using the legend(...) function when no graphic already exists
results in the following error
> legend(x = "topleft", legend = c("a", "b"), fill = rainbow(2))
Error in strwidth(legend, units = "user", cex = cex) :
  plot.new has not been called yet
```

Therefore, to add a legend to your graphics in R, be sure to always create the graphic first, then apply the `legend(...)` function.

Pop quiz

1. An `xlim` value of `c(100, 300)` means which of the following?
   a. Present the data that are not equal to 100 or 300 on the x axis.
   b. Present the data that are equal to 100 or 300 on the x axis.
   c. Present the data that are less than 100 or greater than 300 on the x axis.
   c. Present the data that are between 100 and 300 on the x axis.
2. When should the `legend(...)` function be called?
   a. Before a graphic function is called.
   b. During a graphic function, included as an argument.
   c. After a graphic function.
   d. When a compatible graphic is displayed in the graphic window.

Time for action – creating a scatterplot

A **scatterplot** is a fundamental statistics graphic that can be used to better understand the relationships underlying a dataset. Like descriptive statistics and correlations, scatterplots are especially useful as a precursor to more extensive data analyses, such as linear regression modeling. We can use R to generate scatterplots that depict a single relationship between two variables or the relationships between all of the variables in a dataset. We will practice both of these methods:

1. Use the `plot(...)` function to create a scatterplot depicting a single relationship between two variables:

```
> #create a scatterplot that depicts the relationship between
the number of Shu and Wei soldiers engaged in past fire attacks
> #get the data to be used in the plot
> scatterplotFireWeiSoldiersData <- subsetFire$WeiSoldiers
> scatterplotFireShuSoldiersData <- subsetFire$ShuSoldiers
> #customize the plot
> scatterplotFireSoldiersLabelMain <-
"Soldiers Engaged in Past Fire Attacks"
> scatterplotFireSoldiersLabelX <- "Wei"
> scatterplotFireSoldiersLabelY <- "Shu"
> #use plot(...) to create and display the scatterplot
> plot(x = scatterplotFireWeiSoldiersData,
y = scatterplotFireShuSoldiersData,
main = scatterplotFireSoldiersLabelMain,
xlab = scatterplotFireSoldiersLabelX,
ylab = scatterplotFireSoldiersLabelY)
```

2. Your plot will be displayed in the graphic window, as shown in the following:

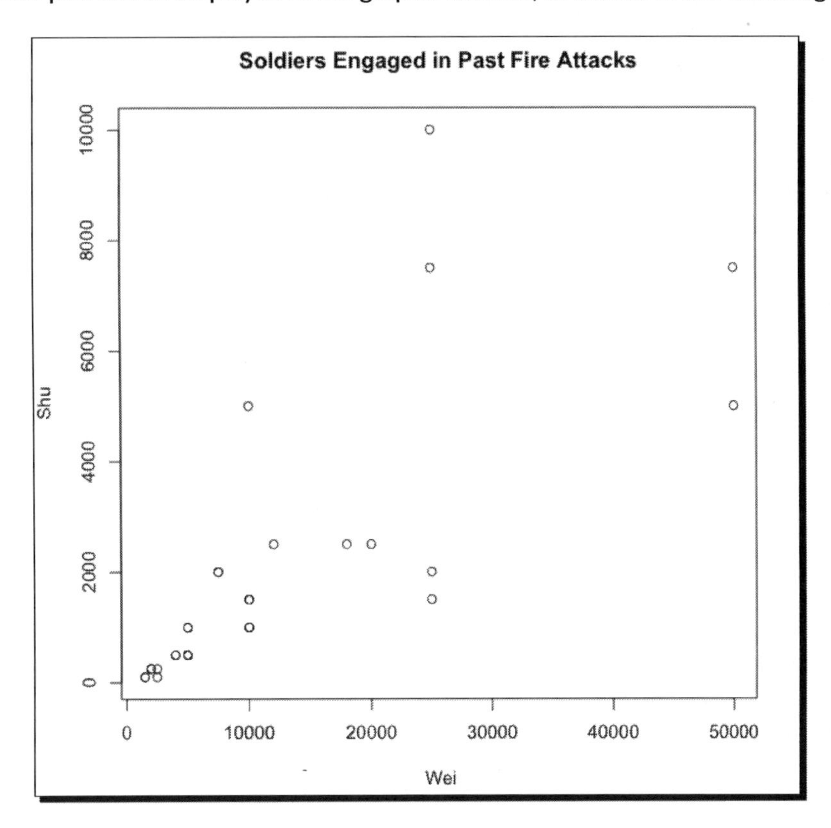

3. Use the `plot(...)` function to simultaneously depict the relationships between all of the variables in the dataset:

```
> #create a scatterplot that depicts the relationships between
all of the variables in our fire attack dataset
> plot(x = subsetFire)
```

4. A grouping of several plots will be displayed in the graphic window:

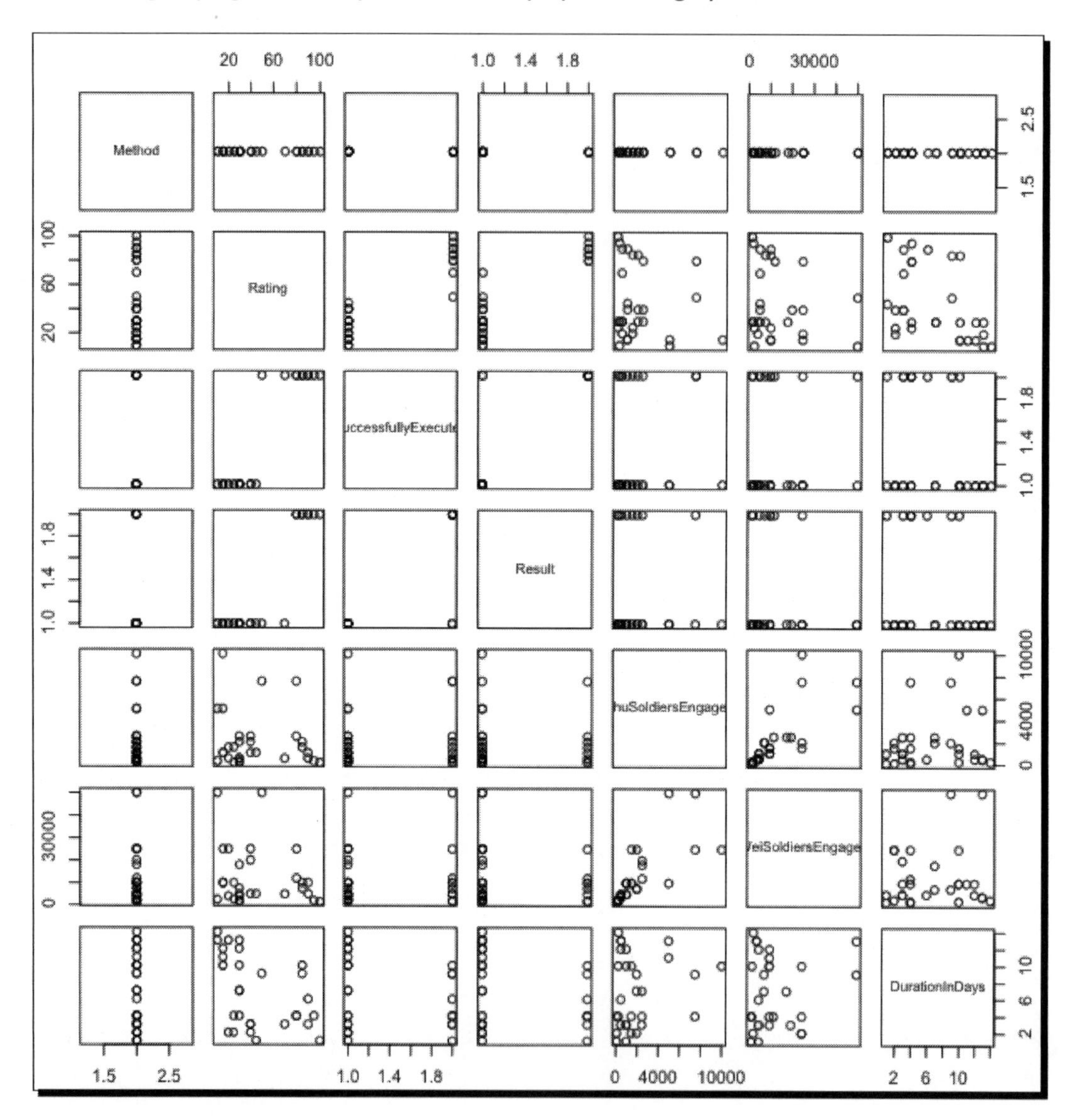

What just happened?

We created two scatterplots using R's `plot(...)` function, one portraying a single relationship and one displaying all of the relationships in our dataset.

Single scatterplot

To plot a single relationship between two variables, use R's `plot(...)` function. The primary arguments for `plot(...)` are:

- `x`: the variable to be plotted on the x axis
- `y`: the variable to be plotted on the y axis

Thus, the simplest form of `plot(...)` contains arguments only for the `x` and `y` variables, and is as shown:

```
plot(x = xVariable, y = yVariable)
```

We used the `plot(...)` function to visualize the relationship between the number of Shu and Wei soldiers involved in past fire attacks. To add relevant text to our graphic, we included the `main`, `xlab`, and `ylab` arguments:

```
> plot(scatterplotFireWeiSoldiersData,
scatterplotFireShuSoldiersData,
main = scatterplotFireSoldiersLabelMain,
xlab = scatterplotFireSoldiersLabelX,
ylab = scatterplotFireSoldiersLabelY)
```

Multiple scatterplots

We also used the `plot(...)` function to simultaneously explore all of the relationships within our dataset. This yielded a graphic that contained a scatterplot for every variable pair. The format for creating this type of scatterplot is:

```
plot(x = dataset)
```

Where `dataset` is a set of data containing multiple variables. For us, the `dataset` argument contained our fire attack data.

```
> plot(x = subsetFire)
```

The resulting plot allowed us to visualize all of the relationships between our variables in a single graphic.

Pop quiz

1. Assume that a and b are data variables. Which of the following best describes the graphic that would result from the following line of code?

```
> plot(x = a, y = b)
```

   a. A scatterplot with a on the x axis and b on the y axis.
   b. A scatterplot with b on the x axis and a on the y axis.
   c. A scatterplot containing all of the relationships in the dataset.
   d. A scatterplot containing none of the relationships in the dataset.

2. Assume that a is a dataset. Which of the following best describes the graphic that would result from the following line of code?

```
> plot(x = a)
```

   a. A scatterplot with a on the x axis.
   b. A scatterplot with a on the y axis.
   c. A scatterplot containing all of the relationships in the dataset.
   d. A scatterplot containing none of the relationships in the dataset.

Time for action – creating a line chart

The ever popular **line chart,** or **line graph**, depicts relationships as continuous series of connected data points. Line charts are particularly useful for visualizing specific values and trends over time. Just as a line chart is an extension of a scatterplot in the non-digital realm, a line chart is created using an extended form of the `plot(...)` function in R. Let us explore how to extend the `plot(...)` function to create line charts in R:

1. Use the `type` argument within the `plot(...)` function to create a line chart that depicts a single relationship between two variables:

```
> #create a line chart that depicts the durations of past fire
attacks
> #get the data to be used in the chart
> lineFireDurationDataX <- c(1:30)
> lineFireDurationDataY <- subsetFire$DurationInDays
> #customize the chart
> lineFireDurationMain <- "Duration of Past Fire Attacks"
> lineFireDurationLabX <- "Battle Number"
> lineFireDurationLabY <- "Duration in Days"
> #use the type argument to connect the data points with a line
```

```
> lineFireDurationType <- "o"
> #use plot(...) to create and display the line chart
> plot(x = lineFireDurationDataX, y = lineFireDurationDataY,
main = lineFireDurationMain, xlab = lineFireDurationLabX,
ylab = lineFireDurationLabY, type = lineFireDurationType)
```

2. Your chart will be displayed in the graphic window, as follows:

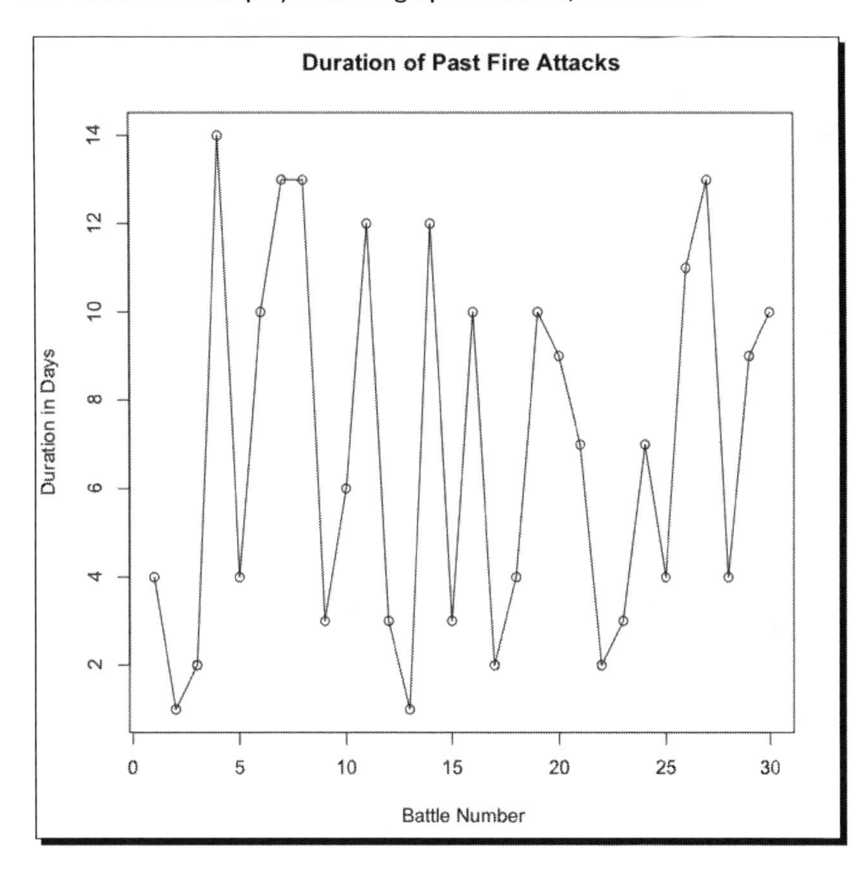

What just happened?

We expanded our use of the `plot(...)` function to generate a line chart and encountered a new data notation in the process. Let us review these features.

type

In the `plot(...)` function, the `type` argument determines what kind of line, if any, should be used to connect a chart's data points. The `type` argument receives one of several character values, all of which are listed as follows:

- `p`: only points are plotted; this is the default value when `type` is undefined
- `l`: only lines are drawn, without any points
- `o`: both lines and points are drawn, with the lines overlapping the points
- `b`: both lines and points are drawn, with the lines broken where they intersect with points
- `c`: only lines are drawn, but they are broken where points would occur
- `s`: only the lines are drawn in step formation; the initial step begins at zero
- `S`: (uppercase) only the lines are drawn in step formation; the final step tails off at the last point
- `h`: vertical lines are drawn to represent each point
- `n`: no points nor lines are drawn

Our chart, which represented the duration of past fire attacks, featured a line that overlapped the plotted points. First, we defined our desired line `type` in an R variable:

```
> lineFireDurationType <- "o"
```

Then the `type` argument was placed within our `plot(...)` function to generate the line chart:

```
> plot(lineFireDurationDataX, lineFireDurationDataY,
main = lineFireDurationMain, xlab = lineFireDurationLabX,
ylab = lineFireDurationLabY,
type = lineFireDurationType)
```

Number-colon-number notation

You may have noticed that we specified a vector for the x-axis data in our `plot(...)` function.

```
> lineFireDurationDataX <- c(1:30)
```

This vector used **number-colon-number notation**. Essentially, this notation has the effect of enumerating a range of values that lie between the number that precedes the colon and the number that follows it. To do so, it adds one to the beginning value until it reaches a final value that is equal to or less than the number that comes after the colon. For example, the code `> 14:21` would yield eight whole numbers, beginning with 14 and ending with 21, as follows:

```
[1] 14 15 16 17 18 19 20 21
```

Furthermore, the code `> 14.2:21` would yield seven values, beginning with 14.2 and ending with 20.2, as follows:

```
[1] 14.2 15.2 16.2 17.2 18.2 19.2 20.2
```

Number-colon-number notation is a useful way to enumerate a series of values without having to type each one individually. It can be used in any circumstance where a series of values is acceptable input into an R function.

Number-colon-number notation can also enumerate values from high to low. For instance, `21:14` would yield a list of values beginning with 21 and ending with 14.

Since we do not have exact dates or other identifying information for our 30 past battles, we simply enumerated the numbers 1 through 30 on the x-axis. This had the effect of assigning a generic identification number to each of our past battles, which in turn allowed us to plot the duration of each battle on the y axis.

Pop quiz

1. Which of the following is the `type` argument capable of?
   a. Drawing a line to connect or replace the points on a scatterplot.
   b. Drawing vertical or step lines.
   c. Drawing no points or lines.
   d. All of the above.
2. What would the following line of code yield in the R console?

   ```
   > 1:50
   ```

   a. A sequence of 50 whole numbers, in order from 1 to 50.
   b. A sequence of 50 whole numbers, in order from 50 to 1.
   c. A sequence of 50 random numbers, in order from 1 to 50.
   d. A sequence of 50 random numbers, in order from 50 to 1.

Time for action – creating a box plot

A useful way to convey a collection of summary statistics in a dataset is through the use of a **box plot**. This type of graph depicts a dataset's minimum and maximum, as well as its lower, median, and upper quartiles in a single diagram. Let us look at how box plots are created in R:

1. Use the `boxplot(...)` function to create a box plot.

```
> #create a box plot that depicts the number of soldiers
required to launch a fire attack
> #get the data to be used in the plot
> boxplotFireShuSoldiersData <- subsetFire$ShuSoldiers
> #customize the plot
> boxPlotFireShuSoldiersLabelMain <- "Number of Soldiers
Required to Launch a Fire Attack"
> boxPlotFireShuSoldiersLabelX <- "Fire Attack Method"
> boxPlotFireShuSoldiersLabelY <- "Number of Soldiers"
> #use boxplot(...) to create and display the box plot
> boxplot(x = boxplotFireShuSoldiersData,
main = boxPlotFireShuSoldiersLabelMain,
xlab = boxPlotFireShuSoldiersLabelX,
ylab = boxPlotFireShuSoldiersLabelY)
```

2. Your plot will be displayed in the graphic window, as shown in the following:

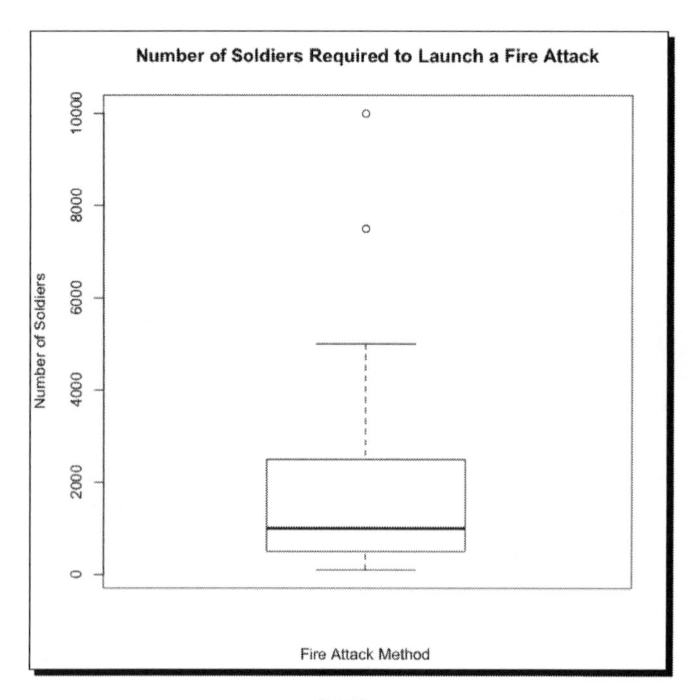

3. Use the `boxplot(...)` function to create a box plot that compares multiple datasets.

```
> #create a box plot that compares the number of soldiers
required across the battle methods
> #get the data formula to be used in the plot
> boxplotAllMethodsShuSoldiersData <- battleHistory$ShuSoldiers
~ battleHistory$Method
> #customize the plot
> boxPlotAllMethodsShuSoldiersLabelMain <- "Number of Soldiers
Required by Battle Method"
> boxPlotAllMethodsShuSoldiersLabelX <- "Battle Method"
> boxPlotAllMethodsShuSoldiersLabelY <- "Number of Soldiers"
> #use boxplot(...) to create and display the box plot
> boxplot(formula = boxplotAllMethodsShuSoldiersData,
main = boxPlotAllMethodsShuSoldiersLabelMain,
xlab = boxPlotAllMethodsShuSoldiersLabelX,
ylab = boxPlotAllMethodsShuSoldiersLabelY)
```

4. Your plot will be displayed in the graphic window, as shown in the following:

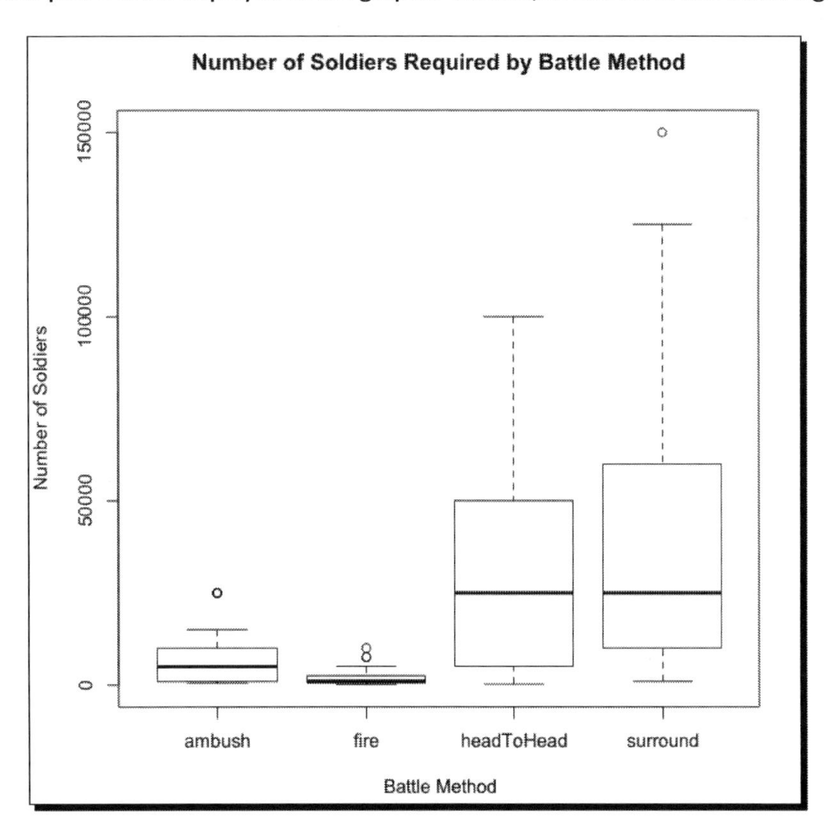

What just happened?

We just created two box plots using R's `boxplot(...)` function, one with a single box and one with multiple boxes.

boxplot(...)

We started by generating a single box plot that was composed of a dataset, main title, and x and y labels. The basic format for a single box plot is as follows:

```
boxplot(x = dataset)
```

The `x` argument contains the data to be plotted. Technically, only `x` is required to create a box plot, although you will often include additional arguments. Our `boxplot(...)` function used the `main`, `xlab`, and `ylab` arguments to display text on the plot, as shown:

```
> boxplot(x = boxplotFireShuSoldiersData,
main = boxPlotFireShuSoldiersLabelMain,
xlab = boxPlotFireShuSoldiersLabelX,
ylab = boxPlotFireShuSoldiersLabelY)
```

Next, we created a multiple box plot that compared the number of Shu soldiers deployed by each battle method. The `main`, `xlab`, and `ylab` arguments remained from our single box plot, however our multiple box plot used the `formula` argument instead of `x`. Here, a formula allows us to break a dataset down into separate groups, thus yielding multiple boxes.

The basic format for a multiple box plot is as follows:

```
boxplot(formula = dataset ~ group)
```

In our case, we took our entire Shu soldier dataset (`battleHistory$ShuSoldiers`) and separated it by battle method (`battleHistory$Method`):

```
> boxplotAllMethodsShuSoldiersData <- battleHistory$ShuSoldiers ~
battleHistory$Method
```

Once incorporated into the `boxplot(...)` function, this formula resulted in a plot that contained four distinct boxes—ambush, fire, head to head, and surround:

```
> boxplot(formula = boxplotAllMethodsShuSoldiersData,
main = boxPlotAllMethodsShuSoldiersLabelMain,
xlab = boxPlotAllMethodsShuSoldiersLabelX,
ylab = boxPlotAllMethodsShuSoldiersLabelY)
```

Pop quiz

1. Which of the following best describes the result of the following code?

```
> boxplot(x = a)
```

a. A single box plot of the a dataset.

b. A single box plot of the x dataset.

c. A multiple box plot of the a dataset that is grouped by x.

d. A multiple box plot of the x dataset that is grouped by a.

2. Which of the following best describes the result of the following code?.

```
> boxplot(formula = a ~ b)
```

a. A single box plot of the a dataset.

b. A single box plot of the b dataset.

c. A multiple box plot of the a dataset that is grouped by b.

d. A multiple box plot of the b dataset that is grouped by a.

Time for action – creating a histogram

A **histogram** displays the frequency with which certain values occur in a dataset. Visually, a histogram looks similar to a bar chart, but it conveys different information. Histograms help us to get an idea of how varied and distributed our data are. Let us begin the histogram making process in R:

1. Use the hist(...) function to create a histogram:

```
> #create a histogram that depicts the frequency distribution
of past fire attack durations
> #get the histogram data
> histFireDurationData <- subsetFire$DurationInDays
> #customize the histogram
> histFireDurationDataMain <- "Duration of Past Fire Attacks"
> histFireDurationLabX <- "Duration in Days"
> histFireDurationLimY <- c(0, 10)
> histFireDurationRainbowColor <-
rainbow(max(histFireDurationData))
> #use hist(...) to create and display the histogram
> hist(x = histFireDurationData,
main = histFireDurationDataMain, xlab = histFireDurationLabX,
ylim = histFireDurationLimY,
col = histFireDurationRainbowColor)
```

2. Your histogram will be displayed in the graphic window, as shown in the following:

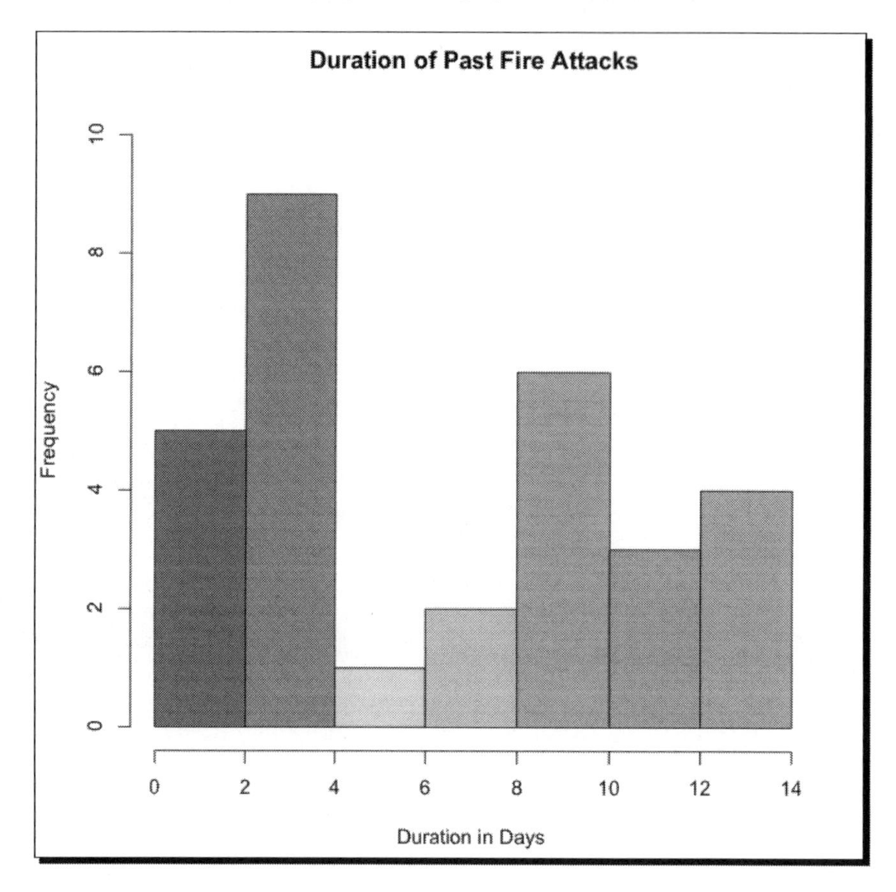

What just happened?

We used the `hist(...)` function to generate a histogram that depicted the frequency distribution of our fire attack duration data.

hist(...)

In its simplest form, the `hist(...)` function is very similar to `boxplot(...)`. At a minimum, it requires only that the data for the chart's columns be defined. A simple function looks like the following:

```
hist(x = dataset)
```

As is true with our other graphics, the `hist(...)` function also receives graphic customization arguments. We rescaled our y-axis with `ylim`, colored our bars with `col`, and added text to our histogram with `main` and `xlab`. Also note that we used the `max(data)` function within the `rainbow(...)` component of our `col` argument to ensure that our histogram would have enough colors to represent each unique value in our dataset:

```
hist(x = histFireDurationData, main = histFireDurationDataMain,
xlab = histFireDurationLabX, ylim = histFireDurationLimY,
col = histFireDurationRainbowColor)
```

Pop quiz

1. Which of the following information are we not capable of deriving from a histogram?
   a. The most and least frequently occurring values in the dataset.
   b. The total number of data points in the dataset.
   c. The minimum and maximum values in the dataset.
   d. The exact value of each data point in the dataset.

Time for action – creating a pie chart

Pie charts are a fast and easy way to visualize a single relationship within a dataset. Let us look at how to create a pie chart in R:

1. Use the `pie(...)` function to create a pie chart:

```
> #create a pie chart that depicts the gold cost of the fire
attack in relation to the total funds allotted to the Shu army
> #get the data to be used in the chart
> #what is the cost of the proposed fire attack?
> functionGoldCost(2500, 225, 7)
[1] 6791.667
> #we already know that 1,000,000 gold has been allotted to
the Shu army
> #therefore our remaining funds after the fire attack would
be 993,208
> #create a vector to hold the values for the chart's slices
> pieFireGoldCostSlices <- c(6792, 993208)
> #use the labels argument to specify the text associated with
each of the chart's slices
> pieFireGoldCostLabels <- c("mission cost", "remaining funds")
> #customize the chart
```

```
> pieFireGoldCostMain <- "Gold Cost of Fire Attack"
> pieFireGoldCostSpecificColors <- c("green", "blue")
> #use pie(...) to create and display the pie chart
> pie(x = pieFireGoldCostSlices,
labels = pieFireGoldCostLabels, main = pieFireGoldCostMain,
col = pieFireGoldCostSpecificColors)
```

2. Your chart will be displayed in the graphic window, similar to the following:

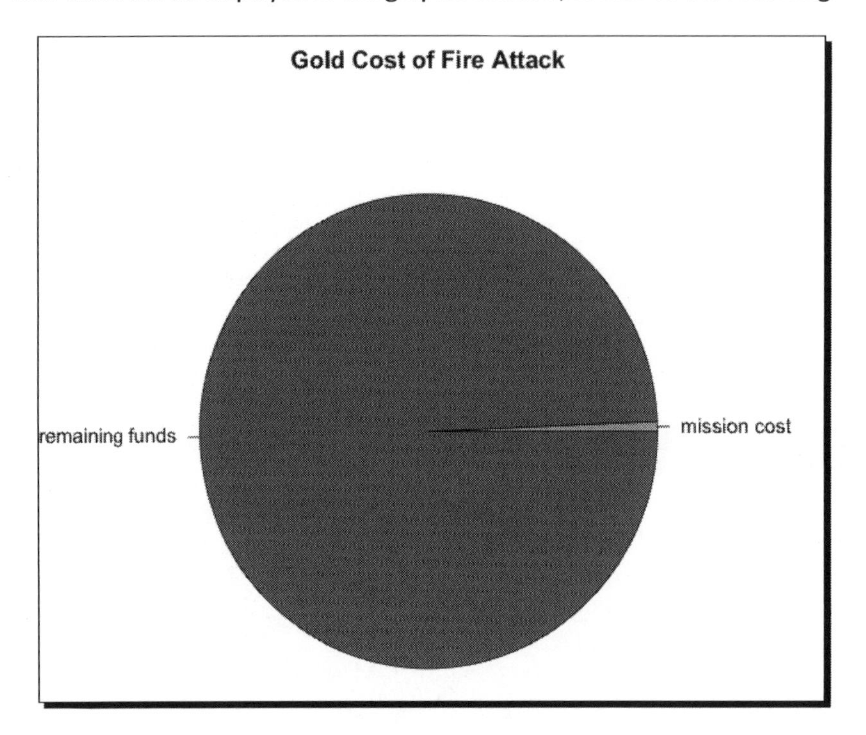

3. Add a legend to the chart, using the following code:

```
> #use the legend(...) function to add a legend to the chart
> legend(x = "bottom", legend = pieFireGoldCostLabels,
fill = pieFireGoldCostSpecificColors)
```

4. Your legend will be added to the existing chart, which will look like the following:

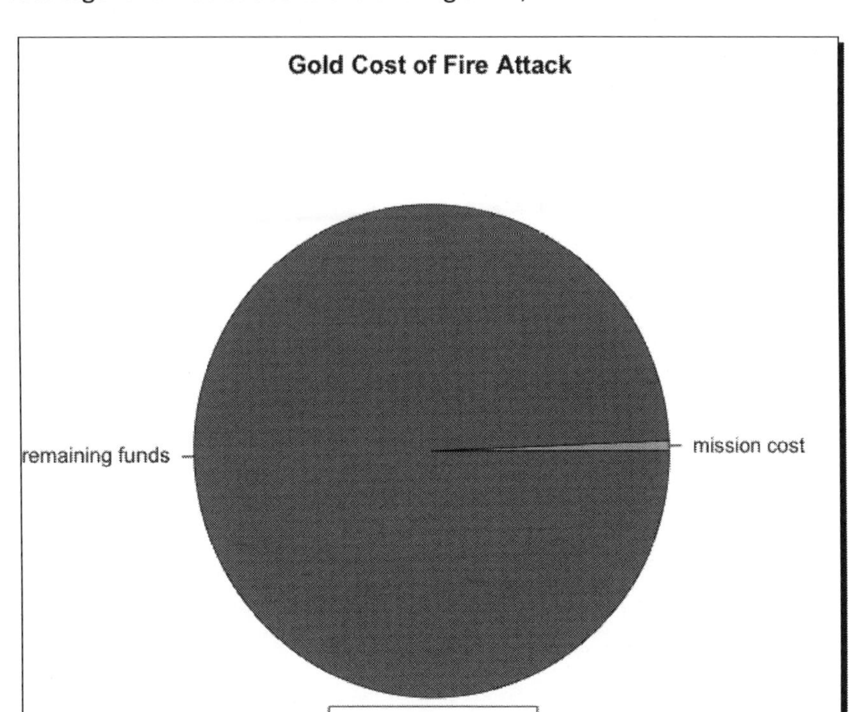

What just happened

We created a pie chart using R's `pie(...)` function and then appended it with a legend. Let us review how pie charts are generated in R.

pie(...)

The primary arguments used in the `pie(...)` function are `x` and `labels`:

- `x`: the numerical values for the pie's slices. These must be nonnegative and input in vector form.
- `labels`: the text labels for the pie's slices. These must be input in vector form.

Consequently, the pie chart function takes on the following basic form:

```
pie(x = sliceData, labels = sliceText)
```

Where `sliceData` and `sliceText` are in vector form.

To create our pie chart, we first calculated the cost information that we wished to display and stored it in a vector variable, like so:

```
> pieFireGoldCostSlices <- c(6792, 993208)
```

Next, we created a vector to hold the text labels for our pie's slices:

```
> pieFireGoldCostLabels <- c("mission cost", "remaining funds")
```

Then we customized our chart with a main title and specific colors, before executing our complete `pie(...)` function:

```
> pie(x = pieFireGoldCostSlices, labels = pieFireGoldCostLabels,
main = pieFireGoldCostMain, col = pieFireGoldCostSpecificColors)
```

Lastly, we added a legend to our chart to further clarify its components:

```
> legend(x = "bottom", legend = pieFireGoldCostLabels,
fill = pieFireGoldCostSpecificColors)
```

Pop quiz

1. In the `pie(...)` function, what do the `x` and `labels` arguments represent?
    a. `labels` represents the slice's numerical values, whereas `x` represents the slice's text labels.
    b. `x` represents the slice's numerical values, whereas `labels` represents the slice's text labels.
    c. `labels` represents the slice's text values, whereas `x` represents the slice's numerical labels.
    d. `x` represents the slice's text values, whereas `labels` represents the slice's numerical labels.

Have a go hero

At this point, you have practiced generating six fundamental R graphics that are critical to data analysis, visualization, and presentation. Use your new R skills to create at least three graphics that will convince the emperor that your battle plan is logistically viable, of benefit the Shu kingdom, and the best choice amongst the competing options. Be sure to employ the customization arguments that we explored in this chapter. Refer back to the individual sections of this chapter for assistance with creating graphics of particular types.

Time for action – exporting graphics

Now that we have created all of these informative graphics, it would be nice to be able to use them for presentations, reports, desktop wallpapers, or a variety of other purposes. Fortunately, R is capable of turning its graphics into digital images that can be used in other applications. Let us look at how to export our graphics for use outside of R:

1. Use one of R's several export functions to convert a graphic into a digital image, it can be done as follows:

```
> #use an export function to save a graphic as a digital image
> #prepare R to export your graphic in one of the following
formats: pdf, png, jpg, tiff, or bmp
> #note that your image will be saved into your R working
directory by default if only a filename is provided
> #otherwise, your image will be saved to the full provided
path
> #optionally, the width and height, in pixels, of the
resulting image can be specified
> #export as pdf
> pdf("myGraphic.pdf", width = 500, height = 500)
> #OR
> #export as png
> png("myGraphic.png", width = 500, height = 500)
> #OR
> #export as jpg
> jpeg("myGraphic.jpg", width = 500, height = 500)
> #OR
> #export as tiff
> tiff("myGraphic.tiff", width = 500, height = 500)
> #OR
> #export as bmp
> bmp("myGraphic.bmp", width = 500, height = 500)
```

2. Create the graphic, as follows:

```
> #create the graphic in R
> #note that your graphic may NOT be displayed in the graphic
window during this process
> #we will use our original fire cost pie chart as an example
> #use pie(...) to create the pie chart
> pie(x = pieFireGoldCostSlices,
```

```
labels = pieFireGoldCostLabels, main = pieFireGoldCostMain,
col = pieFireGoldCostSpecificColors)
> #use the legend(...) function to add a legend to the chart
> legend(x = "bottom", legend = pieFireGoldCostLabels,
fill = pieFireGoldCostSpecificColors)
```

4. Use `dev.off()` to close the current device and export your graphic as a digital image:

```
> #use dev.off() to close the current device and export the
graphic as a digital image
> dev.off()
```

5. Your graphic will be exported. Verify that your digital image has been created.

What just happened?

We just completed the process of exporting an R graphic as a digital image file. Let us detail the three major steps involved in this procedure.

1. **Prepare the graphic device**

   The first step in exporting an R graphic is to prepare the **graphic device**, which is the entity that handles graphics in R. This step requires that a file type for our exported graphic be defined. Optionally, a width and height for the resulting image can also be specified. These can be accomplished through the use of one of several similar functions. These are:

   - `pdf(filename, width, height)`
   - `png(filename, width, height)`
   - `jpeg(filename, width, height)`
   - `tiff(filename, width, height)`
   - `bmp(filename, width, height)`

   Each of these functions prepares the graphic device to export an image associated with its name. For example, the `pdf(filename, width, height)` function will export an image to PDF format. The `filename` argument can contain either a complete path specifying where the image is to be saved or just a filename and extension. If only a name and extension are included, the image will be saved to the R working directory. The `width` and `height` parameters are measured in pixels and receive a single numeric value. For instance, see the following:

```
> pdf("/Users/johnmquick/Desktop/myGraphic.pdf", width = 500,
height = 500)
```

This would export a 500 by 500 pixel PDF image named `myGraphic.pdf` to the given user's desktop. Whereas, look at the following:

```
> pdf("myGraphic.pdf", width = 300, height = 200)
```

This would export a 300 by 200 pixel PDF image named `myGraphic.pdf` to the current working directory.

2. **Create the graphic**

   The second step is to create the graphic in R. This can be done using any of the techniques that we have explored in this chapter. The only difference between this scenario and our previous activities is that we prepared our graphic device prior to creating our graphic. Note that the graphic must be created **after** executing one of the functions provided in the previous step in order to be exported. Also, unlike our other experiences with R visuals, your graphic may **not** be displayed in the graphic window when its function is executed.

3. **Close the graphic device**

   The third and final step is to close the graphics device via the `dev.off()` command. Once `dev.off()` is executed, the graphic will be exported and saved on your computer as a digital image. Afterwards, be sure to check the location that you specified in the first step to verify that your digital image is present and that it was exported properly.

Remembering these three simple steps will allow you to export your R graphics as digital images, thereby allowing them to be used in other applications.

Pop quiz

1. In what order must the three steps of the graphic exportation process proceed?
   a. Create the graphic, prepare the graphic device, close the graphic device.
   b. Close the graphic device, prepare the graphic device, create the graphic.
   c. Prepare the graphic device, close the graphic device, create the graphic.
   d. Prepare the graphic device, create the graphic, close the graphic device.

Have a go hero

Create a custom function named `exportGraphic` that will allow you to save an R graphic as a digital image. Your function should receive five inputs—a filename, a filetype, a width, a height, and a graphics function. For instance, `exportGraphic` should be able to receive the arguments of `test.png`, `png`, `500`, `500`, and `barplot(c(1:10))`, and yield a PNG image of the specified R graphic. Your function should also be able to export an image of any other valid type. Make sure that your custom function follows the process that we used to export our graphics as digital images. Once created, test your `exportGraphic` function to ensure that it works as intended.

Summary

In this chapter, you created several charts, graphs, and plots to communicate your vision and win the approval of the Shu emperor. This process entailed using R's graphical prowess to generate, customize, and export visual representations of your data. At this point, you should be able to:

- Use R to create various charts, graphs, and plots
- Customize your R visuals using colors, lines, and symbols
- Save and export your R visuals

The final stage in preparing for battle is to communicate your strategy to the members of the Shu forces who will execute it. This step requires the simple and clear presentation of precise details. In *Chapter 9*, we will explore the use of detailed custom data visualizations to brief the generals of the Shu army.

9
Briefing the Generals

In Chapter 8, we explored several graphics that can be generated in R. Using these visualizations, you were able to win the favor of the Shu Emperor and receive his approval to carry out your battle plans. Now your focus has turned to the Shu generals, who must be convinced that your plan is worthy of their services. The generals will need to know the details of the attack and how it compares with alternative combat strategies. Recruiting the top generals in the Shu army is critical to the success of your strategy. This challenge calls for clean, detailed, and informative graphics.

We will revisit the charts, graphs, and plots that were created in Chapter 8. To improve their informativeness, clarity, and aesthetics, we will employ new graphics arguments and functions. Specific customization arguments for the different graphics types will be deployed. New graphics functions that add information to visuals will also be explored. We will even work to create our own custom graphics from scratch.

By the end of this chapter, you will be able to do the following:

- Customize several charts, graphs, and plots using arguments specific to each
- Use graphics functions to add information to any visual
- Create custom graphics by building them from the ground up

More charts, graphs, and plots in R

In *Chapter 8*, we customized our graphics using universal arguments that applied to all visuals. However, R graphics often have arguments that apply only to themselves as well. These can be used to make type-specific customizations. We will build upon the graphics that were covered in *Chapter 8* by examining the arguments that are specific to each visual. R also provides graphics functions that can be used to add information to any visual. We will use these to expand our graphics and to experiment with building our own graphics from scratch. The following visuals will be covered in this chapter:

- Bar graphs
- Scatterplots
- Line charts
- Box plots
- Histograms
- Pie charts
- Custom graphics

For demonstration purposes, all of our visuals will communicate information about the fire attack strategy that was used in *Chapter 7*. This strategy entailed deploying 2,500 Shu soldiers for 7 days to execute a fire attack 225 miles away on 10,000 Wei soldiers in Anding. If desired, you are encouraged to substitute your own battle plans into any or all activities for this chapter.

Throughout this chapter, we will modify and build upon the graphics that we created in the previous chapter. All of the necessary variables from that chapter are provided in the `rBeginnersGuide_Ch_09_ReadersCopy.RData` workspace file.

Time for action – customizing a bar chart

To begin, we will expand our *Chapter 8* bar chart using arguments specifically designed for the `barplot(...)` function. We will also become familiar with two different types of bar charts.

1. Open R and set your working directory:

```
> #set the R working directory
> #replace the sample location with one that is relevant to you
> setwd("/Users/johnmquick/rBeginnersGuide/")
```

2. Load the *Chapter 9* workspace. It contains the necessary information for this chapter:

```
> #load the chapter 9 workspace
> load("rBeginnersGuide_Ch_09_ReadersCopy.RData")
```

3. Use the `names`, `width`, and `space` arguments to customize a chart's bars:

```
> #modify the chapter 8 bar chart that compared the mean durations
of the battle methods
> #use the names argument to assign a text label to each bar
> #the names argument receives a vector containing text labels for
each of the chart's bars
> barAllMethodsDurationNames <- c("Fire", "Ambush",
"Head to Head", "Surround")
> #use the width argument to change the width of each bar
> #note that width can be set using a single value for all bars or
by creating a vector to hold a unique value for each bar
> #note that the xlim argument must be defined in order to use the
single value approach
> barAllMethodsDurationLimX <- c(0, 4)
> barAllMethodsDurationWidth <- 0.25
> #use the space argument to change the distance between each bar
> #the space value is a ratio of the average bar width; it
defaults to 0.2
> #note that space can be set using a single value for all bars or
by creating a vector to hold a unique value for each bar
> barAllMethodsDurationSpace <- 2
> #use barplot(...) to create and display the bar chart
> barplot(height = barAllMethodsDurationBars,
main = barAllMethodsDurationLabelMain,
xlab = barAllMethodsDurationLabelX,
ylab = barAllMethodsDurationLabelY,
xlim = barAllMethodsDurationLimX,
ylim = barAllMethodsDurationLimY,
col = barAllMethodsDurationRainbowColors,
names = barAllMethodsDurationNames,
width = barAllMethodsDurationWidth,
space = barAllMethodsDurationSpace)
```

Your chart will be displayed in the graphic window, as shown in the following:

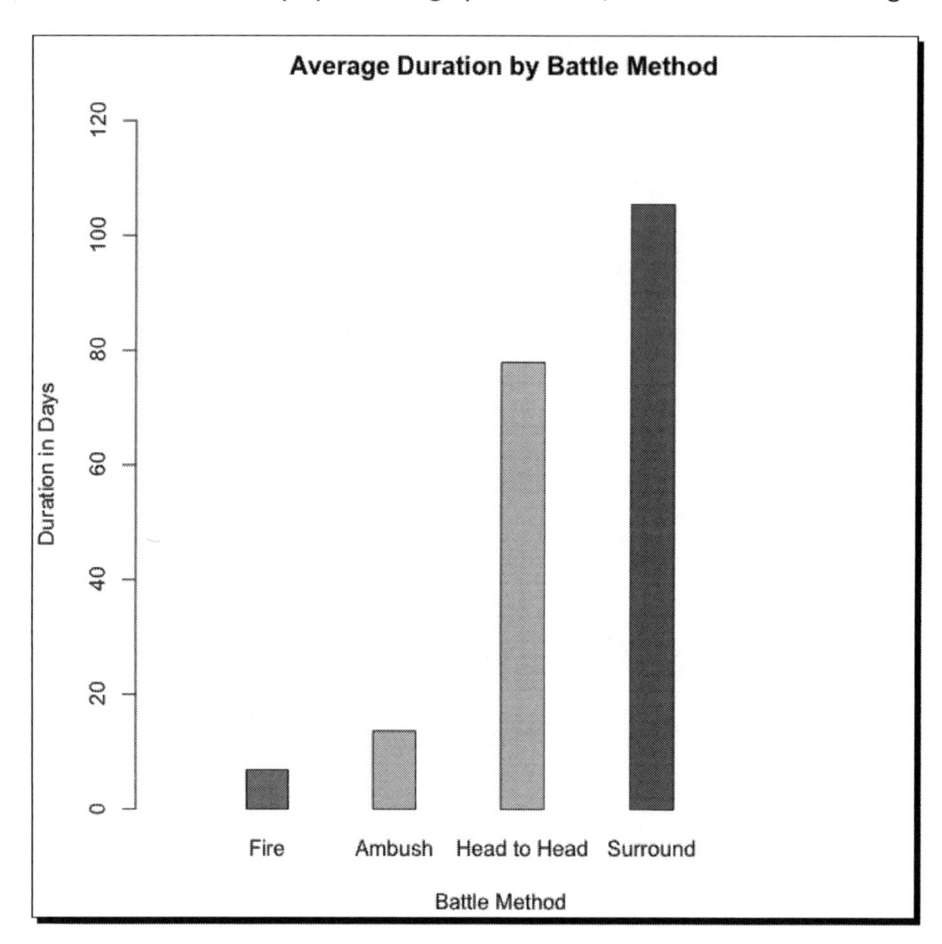

4. Use the `horiz` argument to change the chart's orientation:

```
> #set a bar chart's orientation using the horiz argument
> #if TRUE, the bars will display horizontally
> #if FALSE (default), the bars will display vertically
> barAllMethodsDurationHoriz <- TRUE
> #note that you must reorient the chart for it to display
properly
> #this can be accomplished by switching the values of all
arguments related to the x and y axes
> #use barplot(...) to create and display the bar chart
> barplot(height = barAllMethodsDurationBars,
main = barAllMethodsDurationLabelMain,
xlab = barAllMethodsDurationLabelY,
```

```
ylab = barAllMethodsDurationLabelX,
xlim = barAllMethodsDurationLimY,
ylim = barAllMethodsDurationLimX,
col = barAllMethodsDurationRainbowColors,
names = barAllMethodsDurationNames,
width = barAllMethodsDurationWidth,
space = barAllMethodsDurationSpace,
horiz = barAllMethodsDurationHoriz)
```

Your chart will be displayed in the graphic window, as shown in the following:

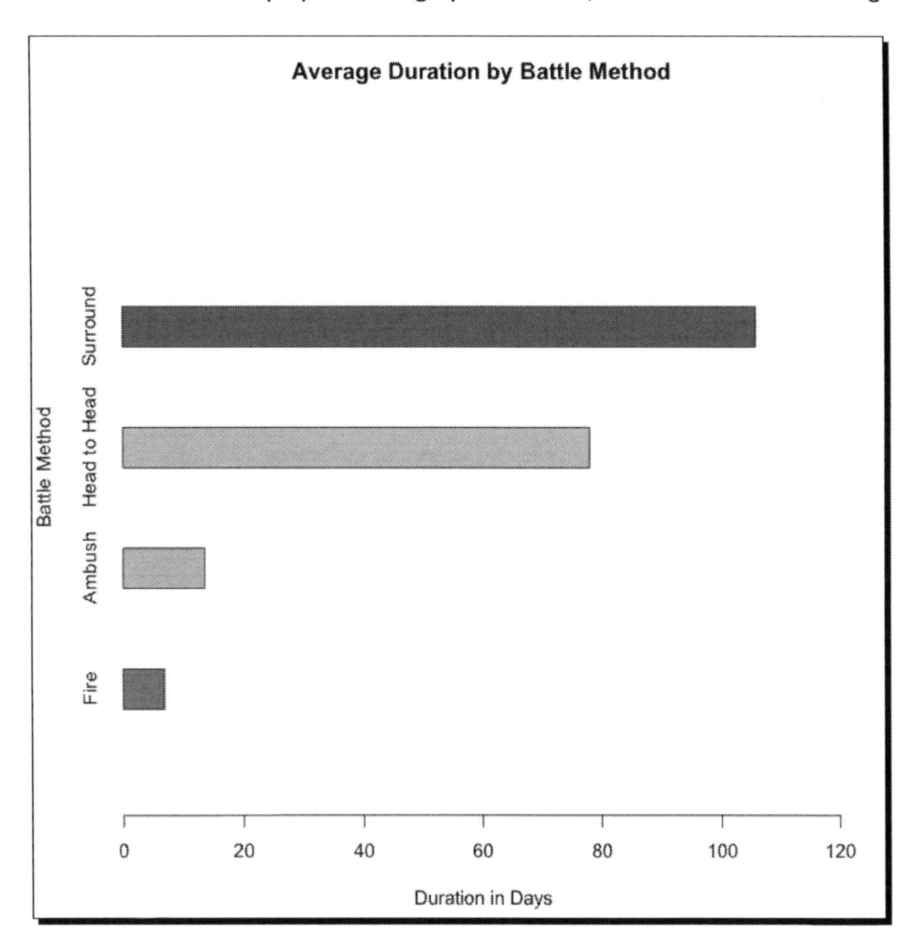

Note that if your bar labels do not all appear along the y-axis, you may want to resize the graphic window. Making your window larger will provide it with enough space to display all of the chart's labels.

5. **Use the `beside` argument to create a stacked bar chart:**

```
> #create a new bar chart to demonstrate the stacking feature
> #create a bar chart that depicts the average number of soldiers
involved in each battle method with stacked bars for the Shu and
Wei forces
> #set the stacking of a chart's bars using the beside argument
> #if TRUE (default), the bars will display next to one another
> #if FALSE, the bars will display atop one another
> barAllMethodsSoldiersBeside <- FALSE
> #note that the bar values must be in matrix form for the beside
argument to take effect
> #calculate the bar values for each method

> #fire
> meanShuSoldiersFire <- mean(subsetFire$ShuSoldiers)
> meanWeiSoldiersFire <- mean(subsetFire$WeiSoldiers)

> #ambush
> meanShuSoldiersAmbush <- mean(subsetAmbush$ShuSoldiers)
> meanWeiSoldiersAmbush <- mean(subsetAmbush$WeiSoldiers)

> #head to head
> meanShuSoldiersHeadToHead <-
mean(subsetHeadToHead$ShuSoldiers)
> meanWeiSoldiersHeadToHead <-
mean(subsetHeadToHead$WeiSoldiers)

> #surround
> meanShuSoldiersSurround <- mean(subsetSurround$ShuSoldiers)
> meanWeiSoldiersSurround <- mean(subsetSurround$WeiSoldiers)

> #put the bar values into matrix form using the matrix(...)
function
> #the matrix should have four columns (one for each method) and
two rows (one for each kingdom)
> #when the chart is created, the rows will be stacked within each
column
> barAllMethodsSoldiersBars <- matrix(c(meanShuSoldiersFire,
meanWeiSoldiersFire, meanShuSoldiersAmbush,
meanWeiSoldiersAmbush, meanShuSoldiersHeadToHead,
meanWeiSoldiersHeadToHead, meanShuSoldiersSurround,
meanWeiSoldiersSurround), 2, 4)
> #customize the chart
> barAllMethodsSoldiersMain <- "Average Number of Soldiers Engaged
```

```
in Battle by Kingdom"
> barAllMethodsSoldiersLabX <- "Battle Method"
> barAllMethodsSoldiersLabY <- "Number of Soldiers"
> barAllMethodsSoldiersNames <- c("Fire", "Ambush",
"Head to Head", "Surround")
> #use barplot(...) to create and display the bar chart
> barplot(height = barAllMethodsSoldiersBars,
main = barAllMethodsSoldiersMain,
xlab = barAllMethodsSoldiersLabX,
ylab = barAllMethodsSoldiersLabY,
names = barAllMethodsSoldiersNames,
beside = barAllMethodsSoldiersBeside)
```

Your chart will be displayed in the graphic window, as follows:

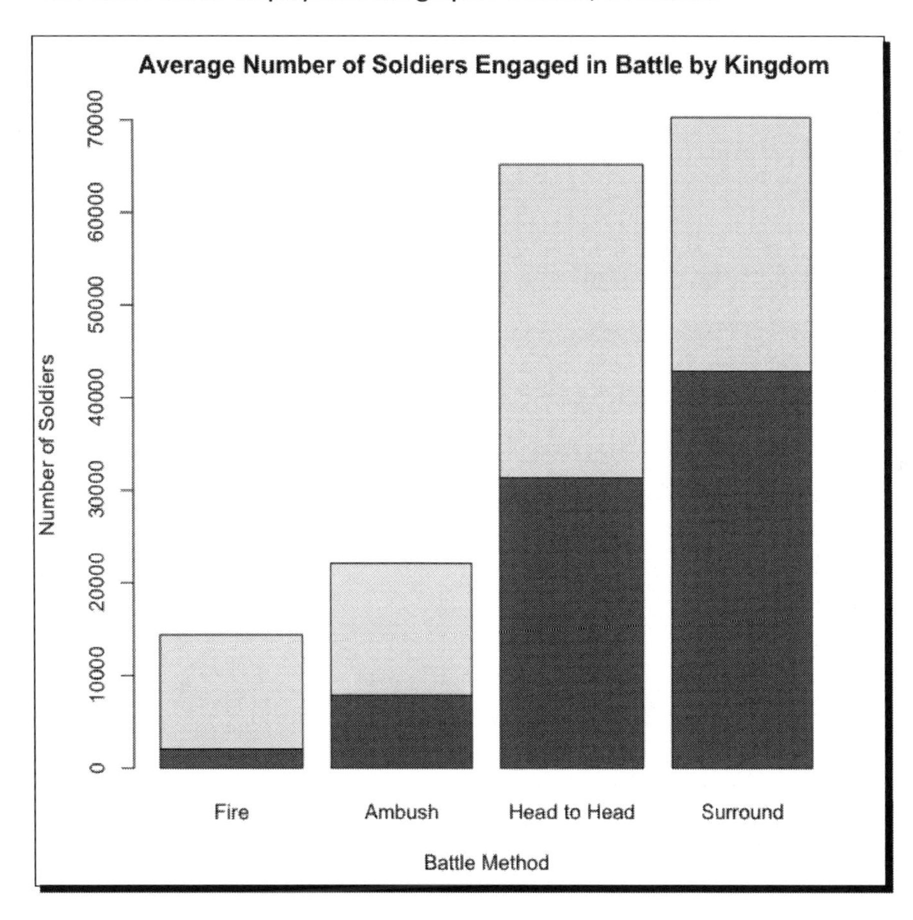

6. **Use the `density` and `angle` arguments to change the shading of the chart's bars:**

```
> #use the density argument to define the thickness of the shaded
lines
> #density receives either a single nonnegative value for all
matrix rows or a vector containing a value for each row
> #density is measured in lines per inch with a default value of
NULL
> barAllMethodsSoldiersDensity <- c(10, 25)
> #angle modifies the angle of the shaded lines
> #angle receives either a single value for all matrix rows or a
vector containing a value for each row
> #angle is measured in degrees
> barAllMethodsSoldiersAngle <- c(45, -45)
> #use barplot(...) to create and display the bar chart
> barplot(height = barAllMethodsSoldiersBars,
main = barAllMethodsSoldiersMain,
xlab = barAllMethodsSoldiersLabX,
ylab = barAllMethodsSoldiersLabY,
names = barAllMethodsSoldiersNames,
beside = barAllMethodsSoldiersBeside,
density = barAllMethodsSoldiersDensity,
angle = barAllMethodsSoldiersAngle)
```

Your chart will be displayed in the graphic window, as shown in the following:

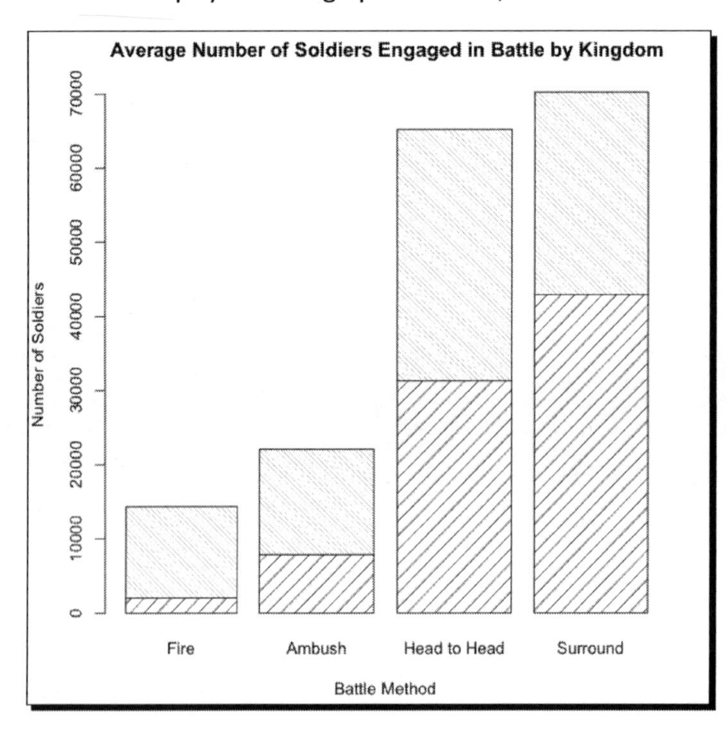

7. Add a legend to the chart:

```
> #add a legend to the stacked bar chart
> #use the x and y arguments to specify the exact location of the legend
> #note that the possible x and y values are determined by the limits of your axes
> #add labels for the Shu and Wei armies
> #incorporate the density and angle arguments from our barplot(...) function
> #use cex to increase the size of the legend
> legend(x = 0.2, y = 70000, legend = c("Shu", "Wei"),
density = barAllMethodsSoldiersDensity,
angle = barAllMethodsSoldiersAngle, cex = 2)
```

Your legend will be added to the existing chart. The final chart looks like the following:

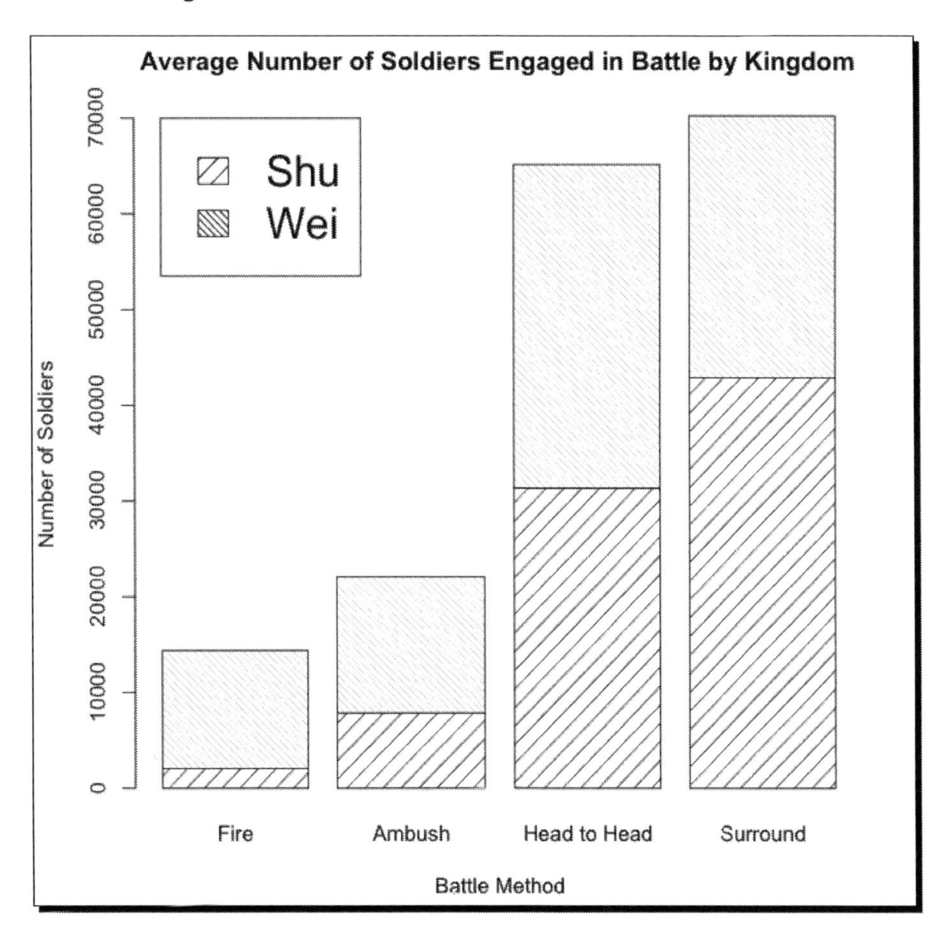

What just happened?

We created vertical, horizontal, and stacked bar charts using the `barplot(...)` function and its custom arguments. We also expanded upon the `legend(...)` function to gain more control over its appearance. Let us reflect upon each of these steps.

names

We started by adding text labels to our bars via the `names` argument. This argument receives a vector containing the text label to be appended to each bar. In our case, the labels consisted of the four battle methods that follow:

```
barAllMethodsSoldiersNames <- c("Fire", "Ambush", "Head to Head",
"Surround")
```

width and space

Then, we looked at two arguments that are unique to the `barplot(...)` function. The `width` argument specifies the thickness of a chart's bars. It can be defined as a single value for all bars or a vector that contains unique values for each bar. Note that if a single value is used, the `xlim` argument must be defined for the `width` argument to take effect. In coordination with `width`, we also employed the `space` argument, which determines the distance between a chart's bars. Like `width`, `space` can be defined as a single value or a vector containing values for each bar. It is measured as a ratio of the average bar width and defaults to a value of 0.2. For example, if the average width of the bars was 5 and the `space` was set to 0.5, then the distance between each bar would be 2.5. For our chart, we chose a width of 0.25 and a space of 2, which had the visual effect of making our bars skinnier and spread farther apart:

```
> barAllMethodsDurationWidth <- 0.25
> barAllMethodsDurationSpace <- 2
```

We chose uniform width and space values for our bars. Had we wanted to set unique values for each bar, such as when weighting the bars according to the number of data points they include, we could have used the following code:

```
> barAllMethodsDurationWidth <- c(0.1, 0.25, 0.5, 0.75)
> barAllMethodsDurationSpace <- c(0.5, 1, 1.5, 2)
```

Lastly, the `names`, `width`, and `space` arguments were incorporated into our overall `barplot(...)` function:

```
> barplot(height = barAllMethodsDurationBars,
main = barAllMethodsDurationLabelMain,
xlab = barAllMethodsDurationLabelX,
ylab = barAllMethodsDurationLabelY,
```

```
xlim = barAllMethodsDurationLimX,
ylim = barAllMethodsDurationLimY,
col = barAllMethodsDurationRainbowColors,
names = barAllMethodsDurationNames,
width = barAllMethodsDurationWidth,
space = barAllMethodsDurationSpace)
```

horiz

To further expand our bar chart, we incorporated the `horiz` argument. This argument allows us to reorient our bars such that they extend horizontally, rather than vertically, across the chart. The `horiz` argument receives either a `TRUE` or `FALSE` value indicating whether the bars should be oriented horizontally. By default, `horiz` is set to `FALSE` and the bars are drawn vertically. We chose to reorient our bars by setting our `horiz` variable to `TRUE`:

```
barAllMethodsDurationHoriz <- TRUE
```

We then included it into our `barplot(...)` function:

```
> barplot(height = barAllMethodsDurationBars,
main = barAllMethodsDurationLabelMain,
xlab = barAllMethodsDurationLabelY,
ylab = barAllMethodsDurationLabelX,
xlim = barAllMethodsDurationLimY,
ylim = barAllMethodsDurationLimX,
col = barAllMethodsDurationRainbowColors,
names = barAllMethodsDurationNames,
width = barAllMethodsDurationWidth,
space = barAllMethodsDurationSpace,
horiz = barAllMethodsDurationHoriz)
```

Note that reorienting the bars of a chart is similar in effect to rotating it by 90 degrees. Therefore, to prevent a misshapen and unreadable graphic, we must also swap all arguments related to the x and y-axes. In our case, that meant exchanging our x-axis and y-axis limits and labels:

```
> barplot(height = barAllMethodsDurationBars,
main = barAllMethodsDurationLabelMain,
xlab = barAllMethodsDurationLabelY,
ylab = barAllMethodsDurationLabelX,
xlim = barAllMethodsDurationLimY,
ylim = barAllMethodsDurationLimX,
col = barAllMethodsDurationRainbowColors,
names = barAllMethodsDurationNames,
width = barAllMethodsDurationWidth,
space = barAllMethodsDurationSpace,
horiz = barAllMethodsDurationHoriz)
```

After swapping these values, the chart displays appropriately. Had we forgotten to make this exchange, we would have ended up with the following graphic:

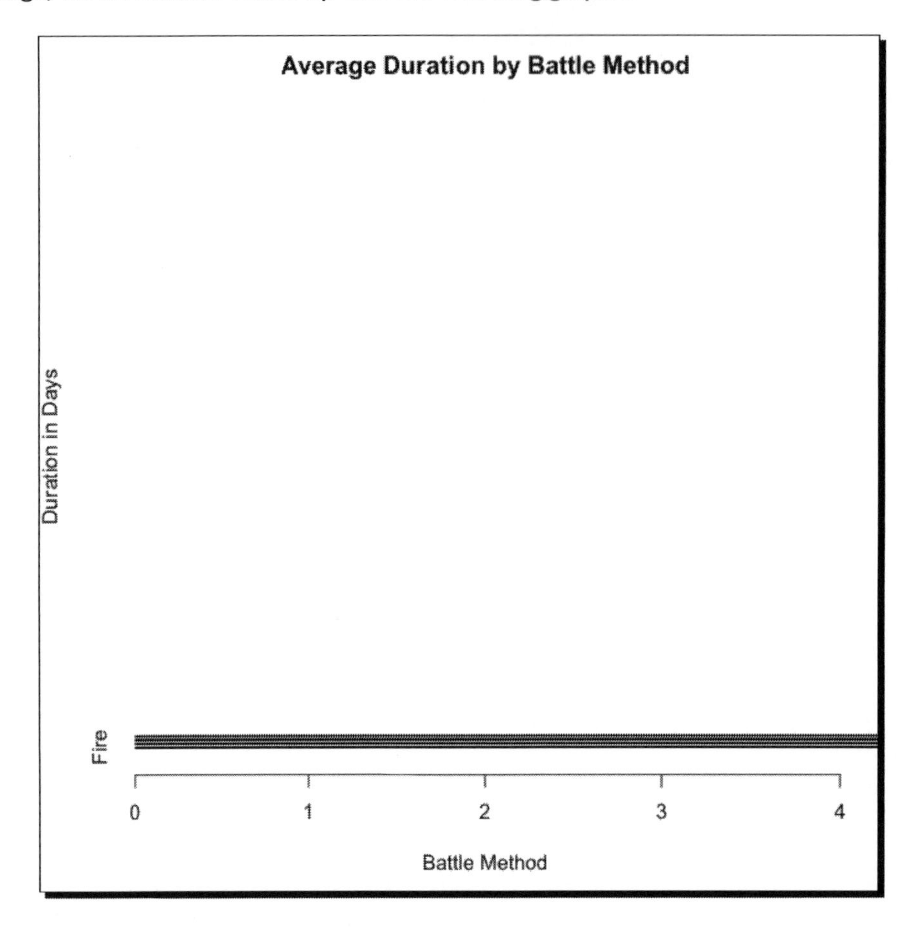

Remember to swap your x-axis and y-axis arguments when making horizontal bar charts.

beside

We then turned to developing a new chart that would make use of the `beside` argument. This argument tells a chart's bars to stack atop one another, rather than stand side by side. Like `horiz`, `beside` accepts a `TRUE` or `FALSE` value. If `TRUE`, the default setting, the bars will display side by side. If `FALSE`, the bars will be stacked. We chose to stack our bars by setting `beside` to `FALSE`.

```
barAllMethodsSoldiersBeside <- FALSE
```

Note that for the beside argument to take effect, the `height` argument must be in matrix form. To organize our data into a matrix, we used the `matrix(...)` function, whose basic format is as follows:

```
matrix(values, rows, columns)
```

Here, `values` is a vector containing the relevant data points, `rows` is the number of rows, and `columns` is the number of columns. Our matrix consisted of eight data points that were organized into two rows (*Shu* and *Wei*) and four columns (*Fire*, *Ambush*, *Head to Head*, and *Surround*). In the final chart, the four bars are formed by stacking the two rows within each column; the following is the code:

```
> barAllMethodsSoldiersBars <- matrix(c(meanShuSoldiersFire,
meanWeiSoldiersFire, meanShuSoldiersAmbush, meanWeiSoldiersAmbush,
meanShuSoldiersHeadToHead, meanWeiSoldiersHeadToHead,
meanShuSoldiersSurround, meanWeiSoldiersSurround), 2, 4)
```

Our stacked bar chart depicted the average number of soldiers that are engaged in each type of battle. By stacking our bars, we were able to specify what proportion of the soldiers came from the Shu and Wei armies. Thus, our chart was able to include more information in the same amount of space:

```
> barplot(height = barAllMethodsSoldiersBars,
main = barAllMethodsSoldiersMain,
xlab = barAllMethodsSoldiersLabX,
ylab = barAllMethodsSoldiersLabY,
names = barAllMethodsSoldiersNames,
beside = barAllMethodsSoldiersBeside)
```

density and angle

After `beside`, we used the `density` and `angle` arguments to define the shading of our bars. The `density` argument defines the closeness of the shaded lines. It receives either a single non-negative value for all matrix rows or a vector that contains values for each row. The `angle` argument specifies the angle at which the shaded lines are to be drawn. It also accepts a single value for all matrix rows or a vector containing values for each row.

Our stacked bar chart used a density of `10` for the Shu row and `25` for the Wei row:

```
> barAllMethodsSoldiersDensity <- c(10, 25)
```

It also featured an angle of `45` for the Shu and `-45` for the Wei:

```
> barAllMethodsSoldiersAngle <- c(45, -45)
```

Hence, you will notice that the shading in the Shu portions of our bars is spread thinner and rises to the upper-right of the chart, whereas the shading in the Wei portions of the chart is thicker and declines towards the lower-right of the chart:

```
> barplot(height = barAllMethodsSoldiersBars,
main = barAllMethodsSoldiersMain,
xlab = barAllMethodsSoldiersLabX,
ylab = barAllMethodsSoldiersLabY,
names = barAllMethodsSoldiersNames,
beside = barAllMethodsSoldiersBeside,
density = barAllMethodsSoldiersDensity,
angle = barAllMethodsSoldiersAngle)
```

legend(...) with density, angle, and cex

In the final step, we added a legend to our chart. A legend is critical to a stacked bar chart, because it indicates the difference between its grouped regions. Our `legend(...)` function expanded upon the legends that we created in the previous chapter. We positioned the legend towards the upper-left side of the chart using the `x` and `y` arguments. We also specified the labels that we wanted to show in the legend (*Shu* and *Wei*). By default, the legend would have displayed the bar names (*Fire*, *Ambush*, *Head to Head*, and *Surround*). Since we needed to display the stacked segments of each bar instead, we had to specifically define them as a vector in the `legend` argument. Next, we incorporated the exact `density` and `angle` arguments from our `barplot(...)` function. This matched the legend's shading to that of our chart. To complete our legend, we used the `cex` argument to multiply its size by `2` times. The `cex` argument accepts a numeric value that indicates how much a legend should be scaled by. Increasing the size of our legend made it easier to read, thus enabling viewers to quickly distinguish between our chart's stacked regions:

```
> legend(x = 0.2, y = 70000, legend = c("Shu", "Wei"),
density = barAllMethodsSoldiersDensity,
angle = barAllMethodsSoldiersAngle, cex = 2)
```

Pop quiz

1. In the `barplot(...)` function, what is the relationship between the `width` and `space` arguments?

    a. `width` sets the distance between the bars, while `space` sets the thickness of the bars.

    b. `width` sets the thickness of the bars, while `space` sets the distance between the bars.

   c. `space` sets the range of the bars on the x-axis, while `width` sets the length of the bars.

   d. `space` sets the length of the bars, while `width` sets the range of the bars on the x-axis.

2. In the `barplot(...)` function, which of the following is **not** critical to note when using the `horiz` argument?

   a. It accepts either a `TRUE` (for horizontal bars) or `FALSE` (for vertical bars) value.

   b. It defaults to `FALSE`.

   c. When `TRUE`, all arguments related to the x and y axes must be swapped for the chart to display properly.

   d. When undefined, the `barplot(...)` function will draw horizontal bars.

3. In the `barplot(...)` function, which of the following is **not** critical to note when using the `beside` argument?

   a. It accepts either a `TRUE` (for adjacent bars) or `FALSE` (for stacked bars) value.

   b. It defaults to `FALSE`.

   c. To take effect, the chart's `height` data must be in matrix form.

   d. When `FALSE`, it is advisable to include a legend with the chart.

Have a go hero

Use your `soldiersByCity` dataset to create a chart that depicts the total number of soldiers in the Shu and Wei armies as two separate bars. Then create a stacked bar chart with the same data, but separate the Shu and Wei bars into distinct sections for each city. Compare these two charts and reflect upon the pros and cons of using each.

Time for action – customizing a scatterplot

Our second look at scatterplots will revolve around customizing data point markers, adding new information to a plot, and creating best fit lines:

1. Customize a scatterplot's point markers using the `pch` and `cex` arguments:

```
> #modify the chapter 8 single scatterplot that depicted the
relationship between the number of Shu and Wei soldiers engaged in
past fire attacks
> #use the pch argument to change the style of the data point
markers
> #pch accepts a whole number value between 0 and 25
> scatterplotFireSoldiersPch <- 2
```

```
> #use the cex argument to change the size of the data point
markers
> #cex accepts a numeric value indicating by how much to scale the
markers
> #cex defaults to value of 1
> scatterplotFireSoldiersCex <- 3
> plot(x = scatterplotFireWeiSoldiersData,
y = scatterplotFireShuSoldiersData,
main = scatterplotFireSoldiersLabelMain,
xlab = scatterplotFireSoldiersLabelX,
ylab = scatterplotFireSoldiersLabelY,
pch = scatterplotFireSoldiersPch,
cex = scatterplotFireSoldiersCex)
```

Your plot will be displayed in the graphic window, as shown in the following:

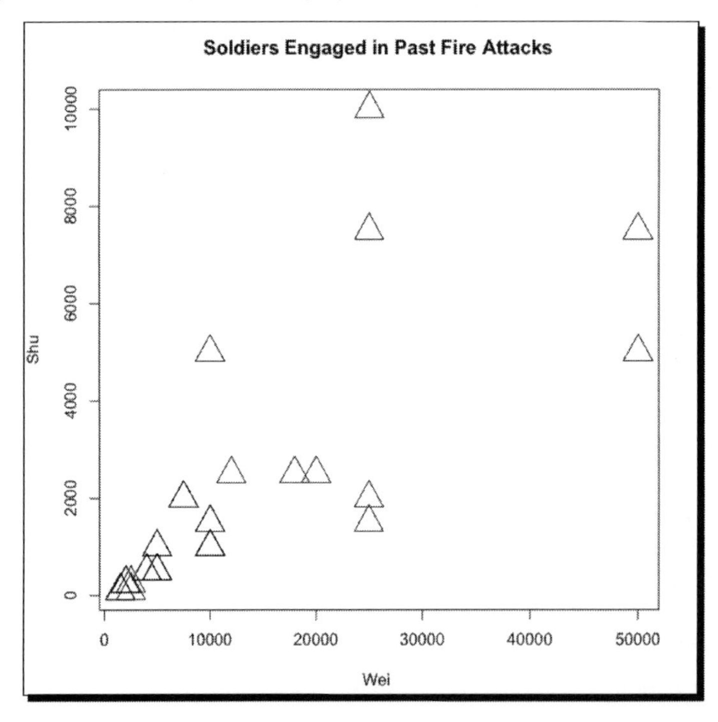

2. Prepare the scatterplot to incorporate additional data:

```
> #prepare the line chart to incorporate data from the other
battle methods
> #modify the chart title
> scatterplotAllMethodsSoldiersMain <-
"Soldiers Engaged by Battle Method"
> #rescale the axes to handle the new data
> scatterplotAllMethodsSoldiersLimX <- c(0, 200000)
```

```
> scatterplotAllMethodsSoldiersLimY <- c(0, 150000)
> #incorporate the col argument to distinguish between the
different battle methods
> scatterplotAllMethodsSoldiersFireCol <- "red"
> #use plot(...) to create and display the revised line chart
> plot(x = scatterplotFireWeiSoldiersData,
y = scatterplotFireShuSoldiersData,
main = scatterplotAllMethodsSoldiersMain,
xlab = scatterplotFireSoldiersLabelX,
ylab = scatterplotFireSoldiersLabelY,
xlim = scatterplotAllMethodsSoldiersLimX,
ylim = scatterplotAllMethodsSoldiersLimY,
col = scatterplotAllMethodsSoldiersFireCol,
pch = scatterplotFireSoldiersPch,
cex = scatterplotFireSoldiersCex)
```

Your scatterplot will be displayed in the graphic window; it will look like the following:

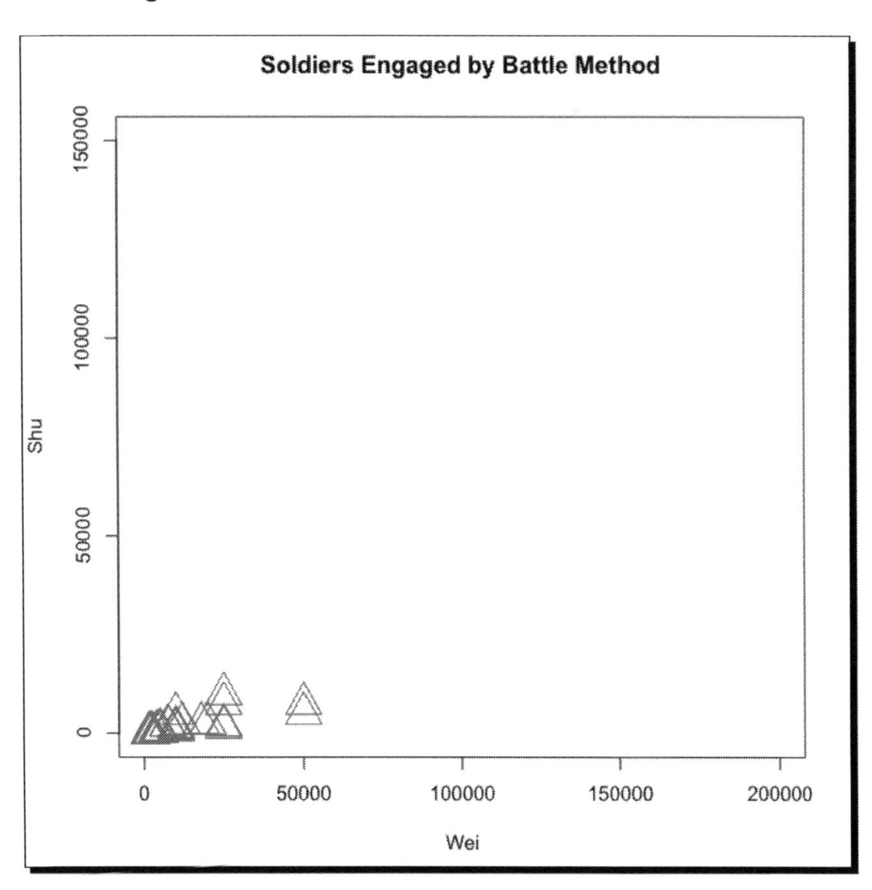

3. **Use the `points(...)` function to add new relationships to the scatterplot:**

```
> #use points(...) to add new relationships to a scatterplot
> #add points representing the three remaining battle methods
> #note that after entering each subsequent function into the
R console, it will be immediately drawn atop your existing
scatterplot

> #ambush
> pointsAmbushDataX <- subsetAmbush$WeiSoldiersEngaged
> pointsAmbushDataY <- subsetAmbush$ShuSoldiersEngaged
> pointsAmbushType <- "p"
> pointsAmbushPch <- 1
> pointsAmbushCex <- 1
> pointsAmbushCol <- "blue"
> points(x = pointsAmbushDataX, y = pointsAmbushDataY,
type = pointsAmbushType, col = pointsAmbushCol,
pch = pointsAmbushPch, cex = pointsAmbushCex)

> #head to head
> pointsHeadToHeadDataX <- subsetHeadToHead$WeiSoldiersEngaged
> pointsHeadToHeadDataY <- subsetHeadToHead$ShuSoldiersEngaged
> pointsHeadToHeadType <- "p"
> pointsHeadToHeadPch <- 3
> pointsHeadToHeadCex <- 1
> pointsHeadToHeadCol <- "darkorange2"
> points(x = pointsHeadToHeadDataX, y = pointsHeadToHeadDataY,
type = pointsHeadToHeadType, col = pointsHeadToHeadCol,
pch = pointsHeadToHeadPch, cex = pointsHeadToHeadCex)

> #surround
> pointsSurroundDataX <- subsetSurround$WeiSoldiersEngaged
> pointsSurroundDataY <- subsetSurround$ShuSoldiersEngaged
> pointsSurroundType <- "p"
> pointsSurroundPch <- 4
> pointsSurroundCex <- 1
> pointsSurroundCol <- "forestgreen"
> points(x = pointsSurroundDataX, y = pointsSurroundDataY,
type = pointsSurroundType, col = pointsSurroundCol,
pch = pointsSurroundPch, cex = pointsSurroundCex)
```

Your points will be added to the existing scatterplot. The scatterplot will look like the following:

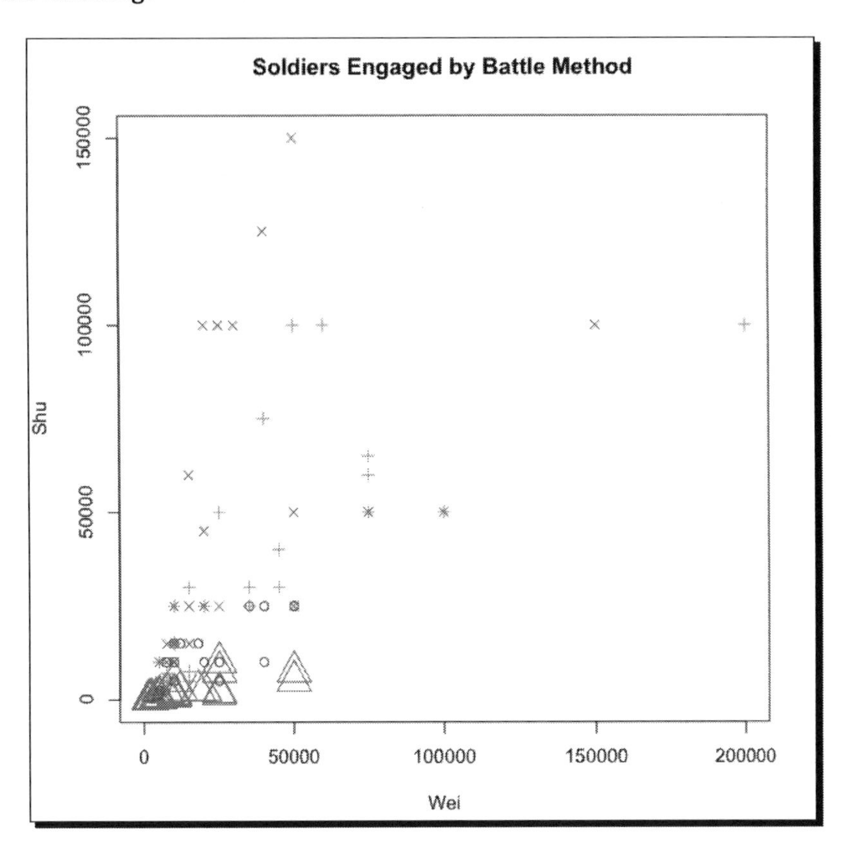

4. Add a legend to the scatterplot.

```
> #add a legend
> #use the x and y arguments to specify the exact location of the
legend
> #add labels for the battle methods
> #add fill colors to match the scatterplot's points
> legend(x = 145000, y = 65000, legend = c("Fire", "Ambush",
"Head to Head", "Surround"), fill = c("red", "blue",
"darkorange2", "forestgreen"))
```

Your legend will be added to the existing scatterplot, which should like like the following:

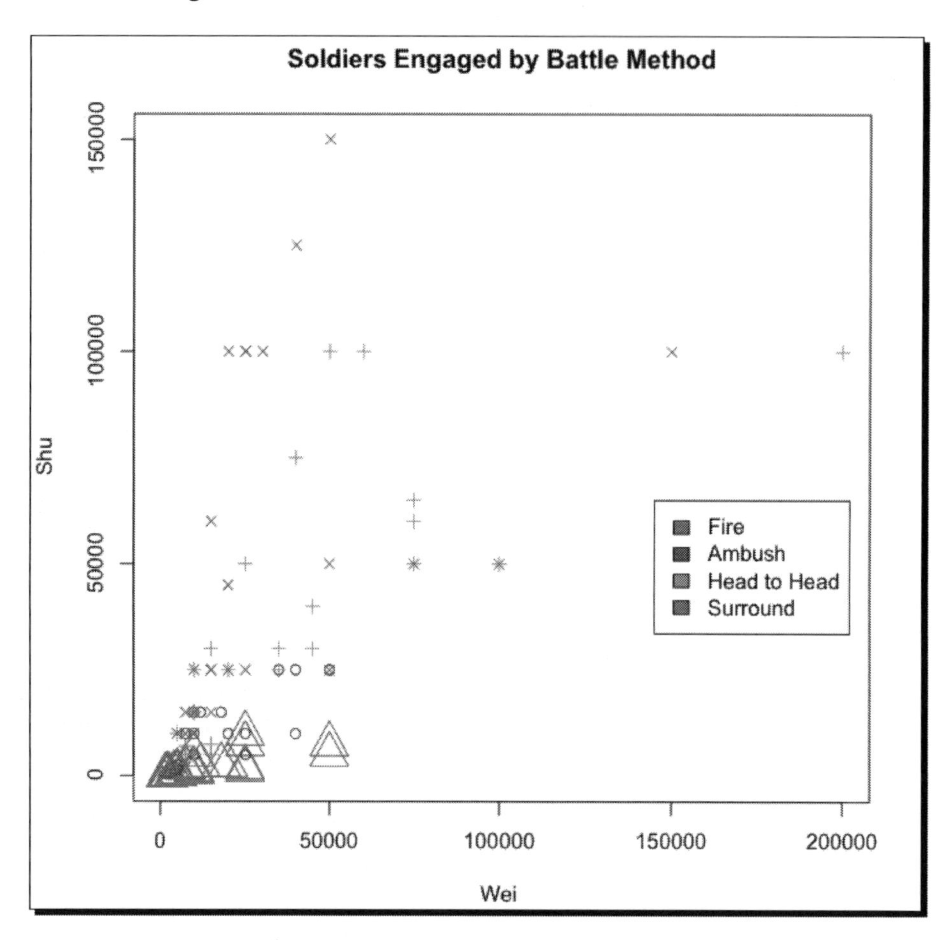

5. Use the `abline(...)` function to add a best fit line to each relationship in the scatterplot.

```
> #add a best fit line using abline(...)
> #the reg argument represents a regression equation
> #reg is defined using the lm(...) function
> #the lty argument defines the style of line to be used
> #as with other graphic functions, the col argument defines a color for the line
> #note that after entering each subsequent function into the R console, it will be immediately drawn atop your existing scatterplot
```

```
> #fire
> scatterplotAllMethodsSoldiersFireLineReg <-
lm(scatterplotFireShuSoldiersData ~
scatterplotFireWeiSoldiersData)
> scatterplotAllMethodsSoldiersFireLty <- "solid"
> #abline(...) will draw a best fit line atop a preexisting plot
> abline(reg = scatterplotAllMethodsSoldiersFireLineReg,
lty = scatterplotAllMethodsSoldiersFireLty,
col = scatterplotAllMethodsSoldiersFireCol)

> #ambush
> scatterplotAllMethodsSoldiersAmbushLineReg <-
lm(pointsAmbushDataY ~ pointsAmbushDataX)
> scatterplotAllMethodsSoldiersAmbushLty <- "dotted"
> #abline(...) will draw a best fit line atop a preexisting plot
> abline(reg = scatterplotAllMethodsSoldiersAmbushLineReg,
lty = scatterplotAllMethodsSoldiersAmbushLty,
col = pointsAmbushCol)

> #head to head
> scatterplotAllMethodsSoldiersHeadToHeadLineReg <-
lm(pointsHeadToHeadDataY ~ pointsHeadToHeadDataX)
> scatterplotAllMethodsSoldiersHeadToHeadLty <- "dotdash"
> #abline(...) will draw a best fit line atop a preexisting plot
> abline(reg = scatterplotAllMethodsSoldiersHeadToHeadLineReg,
lty = scatterplotAllMethodsSoldiersHeadToHeadLty,
col = pointsHeadToHeadCol)

> #surround
> scatterplotAllMethodsSoldiersSurroundLineReg <-
lm(pointsSurroundDataY ~ pointsSurroundDataX)
> scatterplotAllMethodsSoldiersSurroundLty <- "dashed"
> #abline(...) will draw a best fit line atop a preexisting plot
> abline(reg = scatterplotAllMethodsSoldiersSurroundLineReg,
lty = scatterplotAllMethodsSoldiersSurroundLty,
col = pointsSurroundCol)
```

Your best fit lines will be added to the existing scatterplot. The final scatterplot looks like the following:

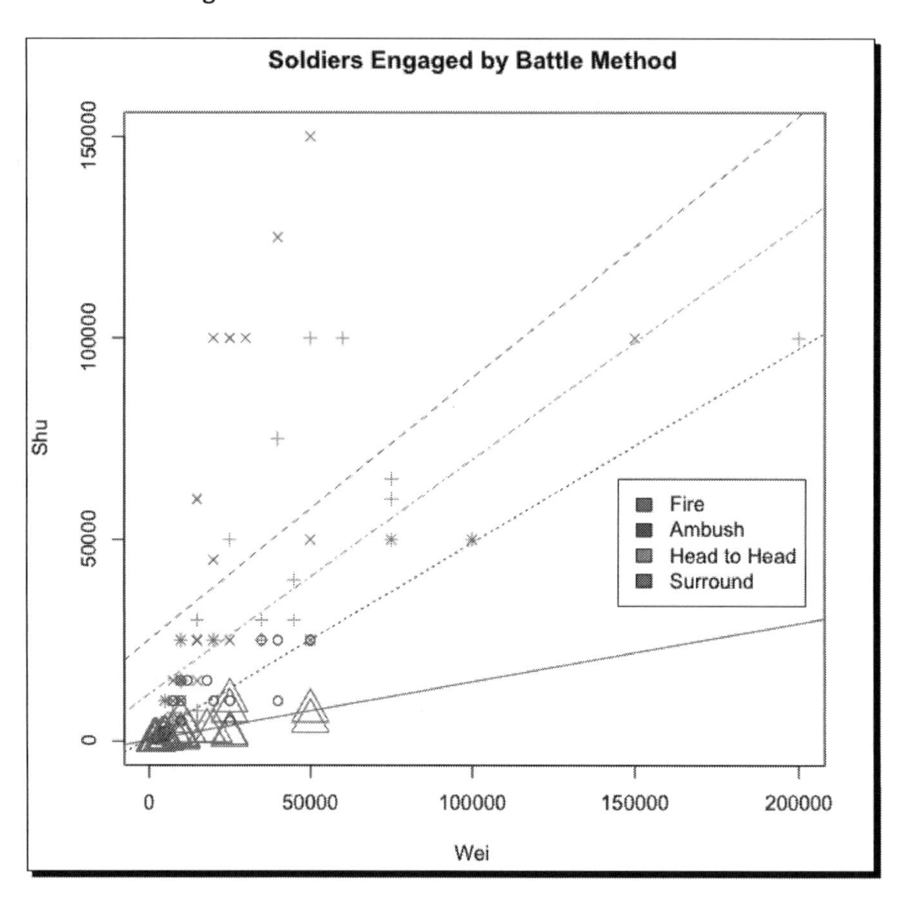

What just happened?

We customized our scatterplot's point markers, then expanded it to include additional data, before adding best fit lines to our graphic. Let us examine these items in greater detail.

pch and cex

We customized the data point markers in our fire attack scatterplot using the `plot(...)` function's `pch` and `cex` arguments. These are defined as follows:

- `pch`: a whole number between 0 and 25, with each value representing a different style of marker, such as a circle, triangle, or square.
- `cex`: a numeric value indicating how much to scale the size of data point markers; 1 by default.

In our case, we used `pch` with the value `2` to apply triangle markers to our data points and then scaled them by three times with `cex` equal to `3`:

```
> scatterplotFireSoldiersPch <- 2
> scatterplotFireSoldiersCex <- 3
```

Thus, we arrived at a plot with large, triangular point markers:

```
> plot(x = scatterplotFireWeiSoldiersData,
y = scatterplotFireShuSoldiersData,
main = scatterplotFireSoldiersLabelMain,
xlab = scatterplotFireSoldiersLabelX,
ylab = scatterplotFireSoldiersLabelY,
pch = scatterplotFireSoldiersPch,
cex = scatterplotFireSoldiersCex)
```

The primary purpose of the `pch` and `cex` arguments is to improve the visual aspects of scatterplots. In tandem, these arguments can generate a wide array of potential data point markers.

You can see a complete list of the markers available for use in the `pch` argument by plotting them with `plot(0:25, pch = 0:25)`.

points(...)

To add new relationships to our scatterplot, we executed the `points(...)` function. This function incorporates additional data points into a plot that is displayed in the graphic window. The primary arguments of the `points(...)` function are:

- `x`: the values to be plotted on the x-axis
- `y`: the values to be plotted on the y-axis
- `type`: the point type; identical to the `type` argument in the `plot(...)` function
- `col`: the point color; identical to the `col` argument in other graphics functions

Thus, the general format for the `points(...)` function is as follows:

```
points(x = xPosition, y = yPosition, type = "type",
col = "colorName")
```

In tandem with these, we also used the `pch` and `cex` arguments in our `points(...)` functions to customize the style and size of our data markers. The `x` and `y` arguments featured the Wei and Shu soldier data for each method:

```
> #ambush
> pointsAmbushDataX <- subsetAmbush$WeiSoldiersEngaged
> pointsAmbushDataY <- subsetAmbush$ShuSoldiersEngaged
> pointsAmbushType <- "p"
> pointsAmbushPch <- 1
> pointsAmbushCex <- 1
> pointsAmbushCol <- "blue"

> #head to head
> pointsHeadToHeadDataX <- subsetHeadToHead$WeiSoldiersEngaged
> pointsHeadToHeadDataY <- subsetHeadToHead$ShuSoldiersEngaged
> pointsHeadToHeadType <- "p"
> pointsHeadToHeadPch <- 3
> pointsHeadToHeadCex <- 1
> pointsHeadToHeadCol <- "darkorange2"

> #surround
> pointsSurroundDataX <- subsetSurround$WeiSoldiersEngaged
> pointsSurroundDataY <- subsetSurround$ShuSoldiersEngaged
> pointsSurroundType <- "p"
> pointsSurroundPch <- 4
> pointsSurroundCex <- 1
> pointsSurroundCol <- "forestgreen"
```

After beginning our scatterplot with fire attack data, we used `points(...)` to plot the soldier data for our ambush, head to head, and surround methods:

```
> #ambush
> points(x = pointsAmbushDataX, y = pointsAmbushDataY,
type = pointsAmbushType, col = pointsAmbushCol,
pch = pointsAmbushPch, cex = pointsAmbushCex)

> #head to head
> points(x = pointsHeadToHeadDataX, y = pointsHeadToHeadDataY,
type = pointsHeadToHeadType, col = pointsHeadToHeadCol,
pch = pointsHeadToHeadPch, cex = pointsHeadToHeadCex)

> #surround
> points(x = pointsSurroundDataX, y = pointsSurroundDataY,
type = pointsSurroundType, col = pointsSurroundCol,
pch = pointsSurroundPch, cex = pointsSurroundCex)
```

> Note that we also redefined the x-axis and y-axis scales with `xlim` and `ylim` prior to adding our new points. This allowed all of our values to display within the bounds of our chart. If we did not rescale the axes, most of our points would fall outside the upper limit of our graph, because the fire attack soldier values are much smaller than in our other battle methods.

legend(...)

We used our familiar `legend(...)` function to add a key that identified the points from each of our battle method datasets. Its title and colors were matched to those of the points in our scatterplot:

```
> legend(x = 145000, y = 65000, legend = c("Fire", "Ambush",
"Head to Head", "Surround"), fill = c("red", "blue", "darkorange2",
"forestgreen"))
```

abline(...)

After completing our scatterplot setup, we added **best fit lines**. Also known as a **regression line**, a best fit line expresses the relationship in a scatterplot as a single, straight line. To accomplish this, the line attempts to orient itself as close as possible to all of the data points. The result is a line that approximates a linear relationship between the variables. In R, we can use the `abline(...)` function to add a best fit line to an existing graphic. In addition to the `col` argument, which we already know about, the primary arguments for `abline(...)` are:

- `reg`: a linear model formula generated by the `lm(...)` function
- `lty`: a text value representing the line style; one of `blank`, `solid`, `dashed`, `dotted`, `dotdash`, `longdash`, or `twodash`

The basic structure of the `abline(...)` function is as follows:

```
abline(reg = lm(y ~ x), lty = "lineType")
```

In our `abline(...)` functions, we used `lty` to define unique line types for each of our battle methods. We also matched our lines' colors to those of our scatterplot's points. Our `reg` arguments used the `lm(...)` function to specify the number of Shu soldiers as our `y` variable and the number of Wei soldiers as our `x` variable:

```
> #fire
> scatterplotAllMethodsSoldiersFireLineReg <-
lm(scatterplotFireShuSoldiersData ~
scatterplotFireWeiSoldiersData)
> scatterplotAllMethodsSoldiersFireLty <- "solid"
```

```
> #ambush
> scatterplotAllMethodsSoldiersAmbushLineReg <-
lm(pointsAmbushDataY ~ pointsAmbushDataX)
> scatterplotAllMethodsSoldiersAmbushLty <- "dotted"

> #head to head
> scatterplotAllMethodsSoldiersHeadToHeadLineReg <-
lm(pointsHeadToHeadDataY ~ pointsHeadToHeadDataX)
> scatterplotAllMethodsSoldiersHeadToHeadLty <- "dotdash"

> #surround
> scatterplotAllMethodsSoldiersSurroundLineReg <-
lm(pointsSurroundDataY ~ pointsSurroundDataX)
> scatterplotAllMethodsSoldiersSurroundLty <- "dashed"
```

The complete `abline(...)` functions incorporated our `reg`, `lty`, and `col` arguments to draw best fit lines for our battle method data:

```
> #fire
> abline(reg = scatterplotAllMethodsSoldiersFireLineReg,
lty = scatterplotAllMethodsSoldiersFireLty,
col = scatterplotAllMethodsSoldiersFireCol)

> #ambush
> abline(reg = scatterplotAllMethodsSoldiersAmbushLineReg,
lty = scatterplotAllMethodsSoldiersAmbushLty,
col = pointsAmbushCol)

> #head to head
> abline(reg = scatterplotAllMethodsSoldiersHeadToHeadLineReg,
lty = scatterplotAllMethodsSoldiersHeadToHeadLty,
col = pointsHeadToHeadCol)

> #surround
> abline(reg = scatterplotAllMethodsSoldiersSurroundLineReg,
lty = scatterplotAllMethodsSoldiersSurroundLty,
col = pointsSurroundCol)
```

A best fit line is useful in gauging whether or not the relationship between two variables is indeed linear. Therefore, it is beneficial to apply when exploring a new dataset. We can also use best fit lines to compare the relationships between related datasets.

In our plot, it is quite clear that the relationship between the numbers of Shu and Wei soldiers engaged is different for different battle methods. For instance, the best fit lines help us to see that in the surround method, the number of Shu soldiers tends to be relatively high compared to the number of Wei soldiers. In contrast, with the fire attack method, the number of Wei soldiers tends to be relatively high compared to the number of Shu soldiers. Using a scatterplot such as this one, along with one or more best fit lines, is still another way to inform our interpretations and understanding of the relationships between our variables. Moreover, using a graphic often helps us to discover things that we cannot see in the raw data alone.

Pop quiz

1. In the `plot(...)` function, what is the relationship between the `pch` and `cex` arguments?
   a. `pch` sets the type of data point marker, while `cex` sets the size of the marker.
   b. `cex` sets the type of data point marker, while `pch` sets the size of the marker.
   c. `pch` sets the number of data point markers, while `cex` sets the style of the markers.
   d. `cex` sets the number of data point markers, while `pch` sets the style of the markers.
2. Which of the following is **not** a benefit of using a scatterplot and best fit line useful to explore the relationship between two variables?
   a. They help us to understand the relationship between the variables.
   b. They inform our interpretation of the relationship between the variables.
   c. They tell us whether the variables will have an interaction effect.
   d. They indicate the linearity of the relationship between the variables.

Have a go hero

Create a scatterplot that depicts the relationship between the execution and rating of past fire attacks. Be sure to use the numeric version of the successful execution variable. Note that since execution is **dichotomous** (containing only two possible values), the resulting plot will look different from the ones we created with our soldier data. Try to interpret the meaning of this graphic. Does it make sense to add a best fit line in this situation?

Now use the `sunflowerplot(...)` function with the same arguments that you just used in the `plot(...)` argument. Try to interpret the meaning of this graphic. Refer back to the raw fire data for help recalling the data contained in the `Rating` and `SuccessfullyExecuted` variables.

Consider the graphics generated by your `plot(...)` and `sunflowerplot(...)` functions. How do these functions differ in the way they portray data?

Time for action – customizing a line chart

To further explore line charts, we will experiment with modifying line widths and adding multiple custom lines to our graphics:

1. Use the `lwd` argument to set the line width:

```
> #modify the chapter 8 single line chart that depicted the
durations of past fire attacks
> #use the lwd argument to set the line width
> #lwd accepts a nonnegative value and defaults to 1
> lineFireDurationWidth <- 3
> #use plot(...) to create and display the line chart
> #recall that a line chart uses the same plot(...) function as a
scatterplot, but with a different type argument
> plot(x = lineFireDurationDataX, y = lineFireDurationDataY,
main = lineFireDurationMain, xlab = lineFireDurationLabX,
ylab = lineFireDurationLabY, type = lineFireDurationType,
lwd = lineFireDurationWidth)
```

Your chart will be displayed in the graphic window, as shown in the following:

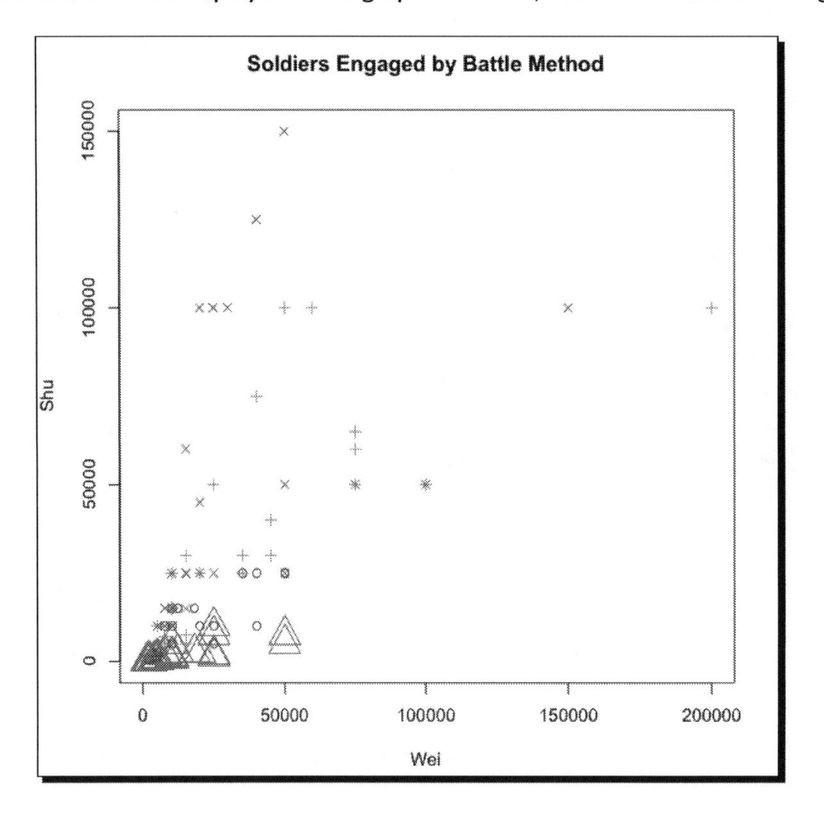

2. Prepare the line chart to incorporate additional data:

```
> #prepare the line chart to incorporate data from the other
battle methods
> #modify the chart title
> lineFireDurationMain = "Duration by Battle Method"
> #rescale the y axis to handle the new data
> lineFireDurationLimY <- c(0, 200)
> #incorporate the col argument to distinguish between the
different battle methods
> lineFireDurationCol <- "red"
> #use plot(...) to create and display the line chart
> plot(x = lineFireDurationDataX, y = lineFireDurationDataY,
main = lineFireDurationMain, xlab = lineFireDurationLabX,
ylab = lineFireDurationLabY, ylim = lineFireDurationLimY,
type = lineFireDurationType, lwd = lineFireDurationWidth,
col = lineFireDurationCol)
```

Your chart will be displayed in the graphic window as shown:

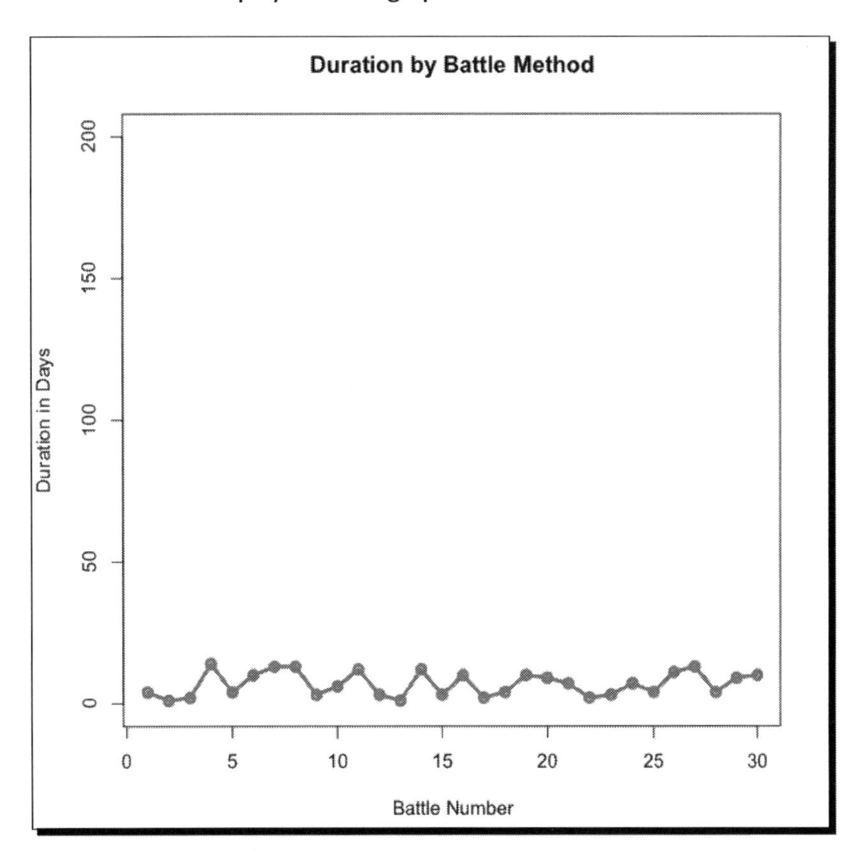

3. **Use the `lines(...)` function to add new relationships to the line chart:**

```
> #use lines(...) to add new relationships to a line chart
> #add lines representing the three remaining battle methods to
the chart
> #note that after entering each subsequent function into the R
console, it will be immediately drawn atop your existing line
chart

> #ambush
> lineAmbushDataY <- subsetAmbush$DurationInDays
> lineAmbushWidth <- 1
> lineAmbushCol <- "blue"
> lines(x = lineFireDurationDataX, y = lineAmbushDataY,
type = lineFireDurationType, lwd = lineAmbushWidth,
col = lineAmbushCol)

> #head to head
> lineHeadToHeadDataY <- subsetHeadToHead$DurationInDays
> lineHeadToHeadWidth <- 1
> lineHeadToHeadCol <- "darkorange2"
> lines(x = lineFireDurationDataX, y = lineHeadToHeadDataY,
type = lineFireDurationType, lwd = lineHeadToHeadWidth,
col = lineHeadToHeadCol)

> #surround
> lineSurroundDataY <- subsetSurround$DurationInDays
> lineSurroundWidth <- 1
> lineSurroundCol <- "forestgreen"
> lines(x = lineFireDurationDataX, y = lineSurroundDataY,
type = lineFireDurationType, lwd = lineSurroundWidth,
col = lineSurroundCol)
```

Your lines will be added to the existing chart, as shown in the following:

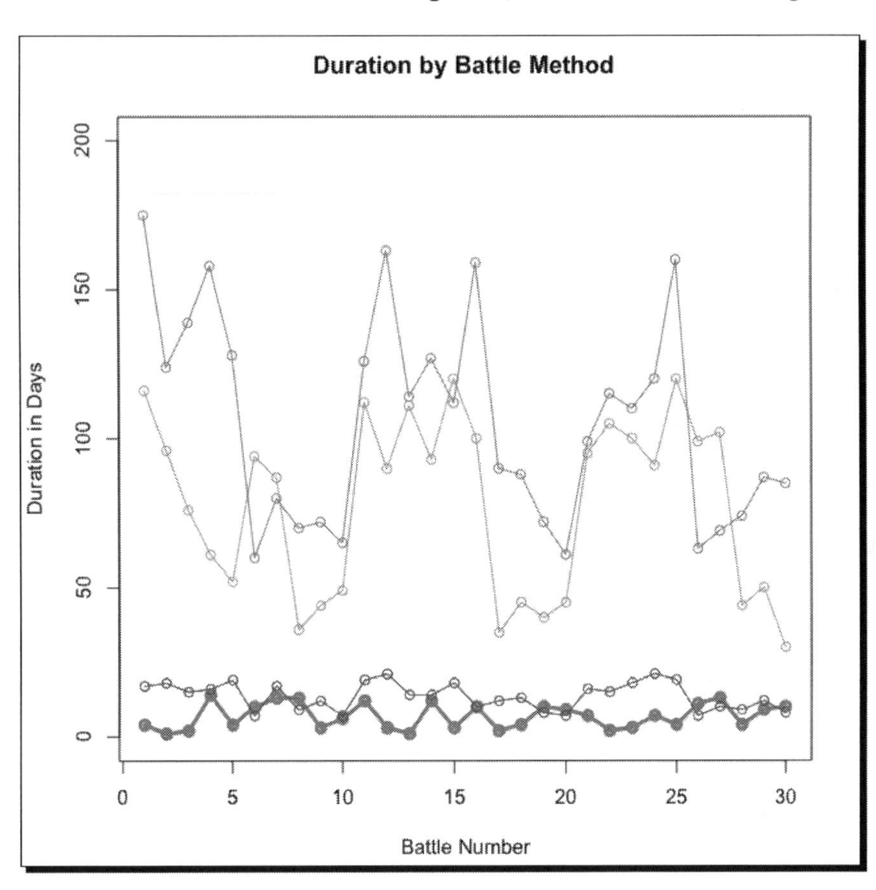

4. Add a legend to the chart in the following way:

```
> #add a legend to our line chart
> #use the x and y arguments to specify the exact location of the
legend
> #add labels for the battle methods
> #add fill colors to match the chart's lines
> legend(x = 23, y = 210, legend = c("Fire", "Ambush",
"Head to Head", "Surround"), fill = c("red", "blue",
"darkorange2", "forestgreen"))
```

5. Your legend will be added to the existing chart; the final chart looks like the following:

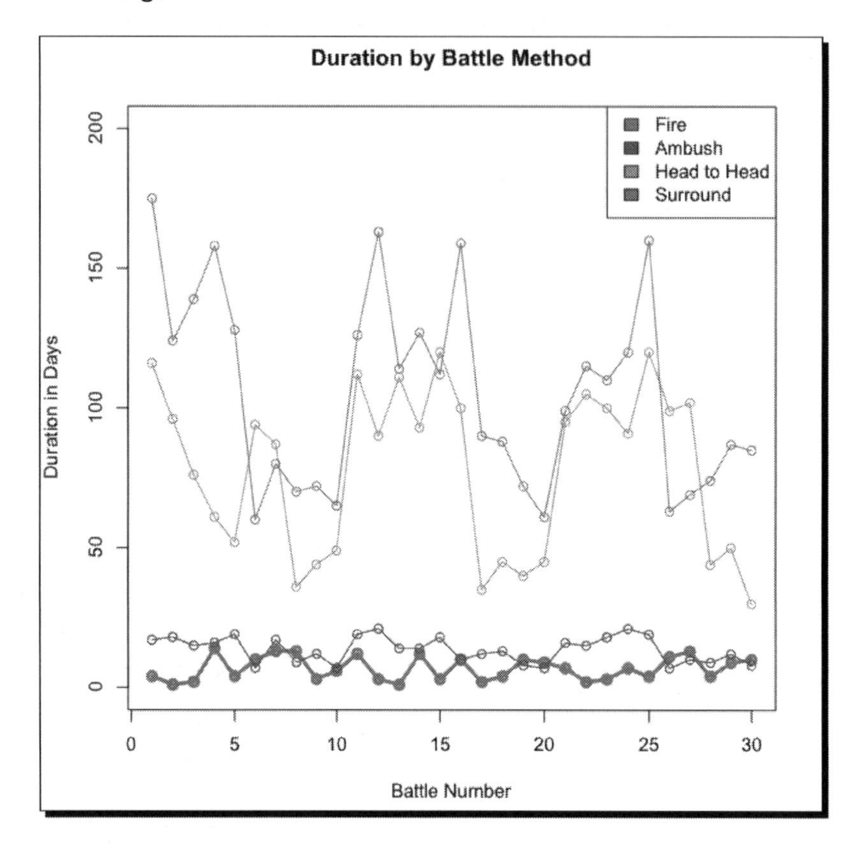

What just happened?

We expanded our use of the `plot(...)` function to generate a line chart with a specific line width. Then, we worked to add additional lines to our chart for the purpose of portraying multiple relationships. We also incorporated a legend to make our chart more legible.
Let us review these techniques.

lwd

We specified the width of our chart's line using the `lwd` argument. This argument has a default value of 1 and can receive any number greater than zero. In most cases, you will want to use values between 1 and 3. Both our one-line and multiline charts used a `lwd` value of 3 to emphasize the fire attack duration data by thickening its line.

```
> lineFireDurationWidth <- 3
```

The `lwd` argument was seamlessly integrated into our `plot(...)` function:

```
> plot(x = lineFireDurationDataX, y = lineFireDurationDataY,
main = lineFireDurationMain, xlab = lineFireDurationLabX,
ylab = lineFireDurationLabY, type = lineFireDurationType,
lwd = lineFireDurationWidth)
```

Note that the `lwd` argument can be used to modify the line thickness of data markers. For example, using a `lwd` of 3 in a scatterplot would yield points with thicker markers. The `lwd` argument can also be used within the `abline(...)` function to alter a best fit line.

lines(...)

To add new relationships to our multiline chart, we employed the `lines(...)` function. This function is used to draw additional lines on the chart that is displayed in the graphic window. The primary arguments of the `lines(...)` function are:

- `x`: the values to be plotted on the x-axis
- `y`: the values to be plotted on the y-axis
- `type`: the line type; identical to the `type` argument in the `plot(...)` function
- `col`: the line color; identical to the `col` argument in other graphics functions

Thus, the general format for the `lines(...)` function is as follows:

```
lines(x = xPosition, y = yPosition, type = "type",
col = "colorName")
```

After generating our chart with only fire attack data, we used `lines(...)` to graph the duration values for our ambush, head to head, and surround methods. For these lines, we used a more subtle `lwd` value of 1 and custom colors to differentiate them from one another.

```
> #ambush
> lineAmbushWidth <- 1
> lineAmbushCol <- "blue"

> #head to head
> lineHeadToHeadWidth <- 1
> lineHeadToHeadCol <- "darkorange2"

> #surround
> lineSurroundWidth <- 1
> lineSurroundCol <- "forestgreen"
```

A unique `line(...)` function for battle method was executed to add its data to the line chart:

```
> #ambush
> lines(x = lineFireDurationDataX, y = lineAmbushDataY,
type = lineFireDurationType, lwd = lineAmbushWidth,
col = lineAmbushCol)

> #head to head
> lines(x = lineFireDurationDataX, y = lineHeadToHeadDataY,
type = lineFireDurationType, lwd = lineHeadToHeadWidth,
col = lineHeadToHeadCol)

> #surround
> lines(x = lineFireDurationDataX, y = lineSurroundDataY,
type = lineFireDurationType, lwd = lineSurroundWidth,
col = lineSurroundCol)
```

Note that we also redefined the y-axis scale with `ylim` prior to adding our new lines. This is necessary, because it allows all of our values to display within the bounds of our chart. If we did not rescale the y-axis, most of our points would fall outside the upper limit of our graph. This is because the fire attack duration values are much smaller than in our other battle methods.

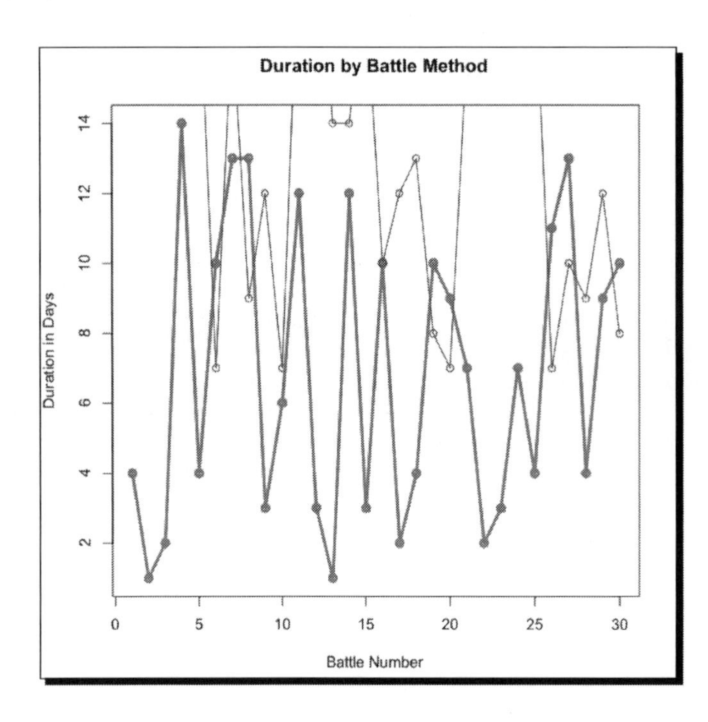

When adding new relationships to a graphic, remember to adjust your axes accordingly to ensure that all data are represented.

legend(...)

Once again, we added a legend to our chart in order to identify each line. We used the already familiar `legend(...)` function to do so, making sure to match the legend's title and colors to those of the lines on our chart:

```
> legend(x = 23, y = 210, legend = c("Fire", "Ambush",
"Head to Head", "Surround"), fill = c("red", "blue", "darkorange2",
"forestgreen"))
```

Pop quiz

1. In the `plot(...)` function, which of the following is **not** true of the `lwd` argument?
    a. One or more of a chart's lines can have a unique `lwd` value.
    b. The `lwd` argument defaults to a value of 1.
    c. The `lwd` argument accepts a nonnegative numeric value.
    d. To take effect, the `lty` argument must be defined.
2. When using the `lines(...)` function to add new lines to a chart, which of the following is **not** a true statement?
    a. One or more lines can be added to a single chart.
    b. The widths of a chart's lines can be different.
    c. To display a new line, the chart's data must be in matrix form.
    d. The x or y axis may need to be rescaled to properly portray a new line.

Have a go hero

Create a multiline chart that portrays the number of Shu soldiers engaged in all instances of each battle method. You should have a line for each battle method. Be sure to experiment with the `type` and `lwd` arguments, as well as the `lines(...)` function, to witness the different line chart styles that can be generated in R. Once your graph is complete, remember to add a legend that identifies each line.

Time for action – customizing a box plot

In learning to customize box plots, we will alter whisker lengths and create custom axes for our graphics.

1. Use the `range` argument to alter the whisker length of each box:

```
> #modify the chapter 8 multiple box plot that that compares the
number of Shu soldiers required across the battle methods
> #rescale the y axis to best display the new range
> boxPlotAllMethodsShuSoldiersLimY <- c(0, 100000)
> #use the range argument to alter the whisker length of each box
> #use range = 0 to extend the whiskers to the most extreme points
> #use range > 0 to extend the whisker to a value of n times the
interquartile range
> #here, limit the whisker range to 1 times the interquartile
range
> boxPlotAllMethodsShuSoldiersRange <- 1
> #use boxplot(...) to create and display the revised line chart
> boxplot(formula = boxplotAllMethodsShuSoldiersData,
main = boxPlotAllMethodsShuSoldiersLabelMain,
xlab = boxPlotAllMethodsShuSoldiersLabelX,
ylab = boxPlotAllMethodsShuSoldiersLabelY,
ylim = boxPlotAllMethodsShuSoldiersLimY,
range = boxPlotAllMethodsShuSoldiersRange)
```

Your plot will be displayed in the graphic window. Note the rescaling of the y-axis and the change in whisker length for the boxes:

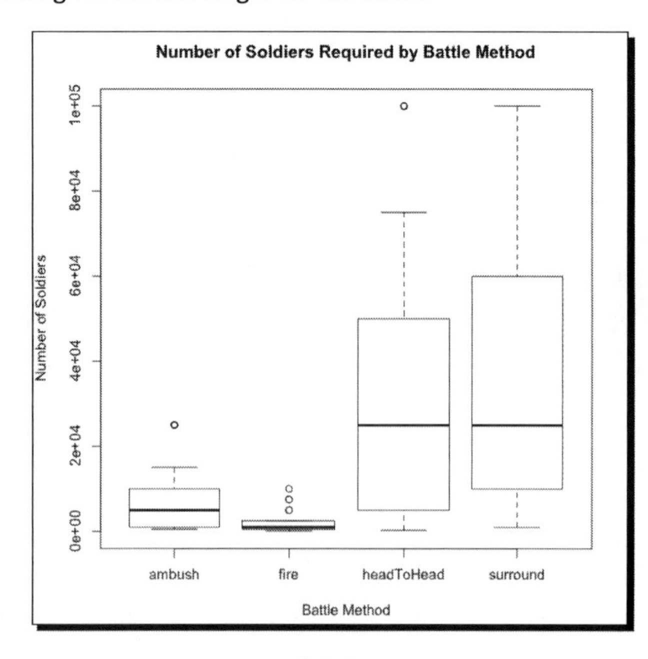

2. Prepare to create custom axes by hiding your box plot's default axes:

```
> #hide the box plot's default axes
> #redraw the box plot using the xaxt and yaxt arguments to hide
the axes
> boxplotAllMethodsShuSoldiersAxtX = "n"
> boxplotAllMethodsShuSoldiersAxtY = "n"
> #use boxplot(...) to create a display the box plot
> #your box plot will have no labels or tick marks on the x and y
axes
> boxplot(formula = boxplotAllMethodsShuSoldiersData,
main = boxPlotAllMethodsShuSoldiersLabelMain,
xlab = boxPlotAllMethodsShuSoldiersLabelX,
ylab = boxPlotAllMethodsShuSoldiersLabelY,
ylim = boxPlotAllMethodsShuSoldiersLimY,
range = boxPlotAllMethodsShuSoldiersRange,
xaxt = boxplotAllMethodsShuSoldiersAxtX,
yaxt = boxplotAllMethodsShuSoldiersAxtY)
```

Your plot will be displayed in the graphic window. Note the lack of x-axis and y-axis labels and tick marks:

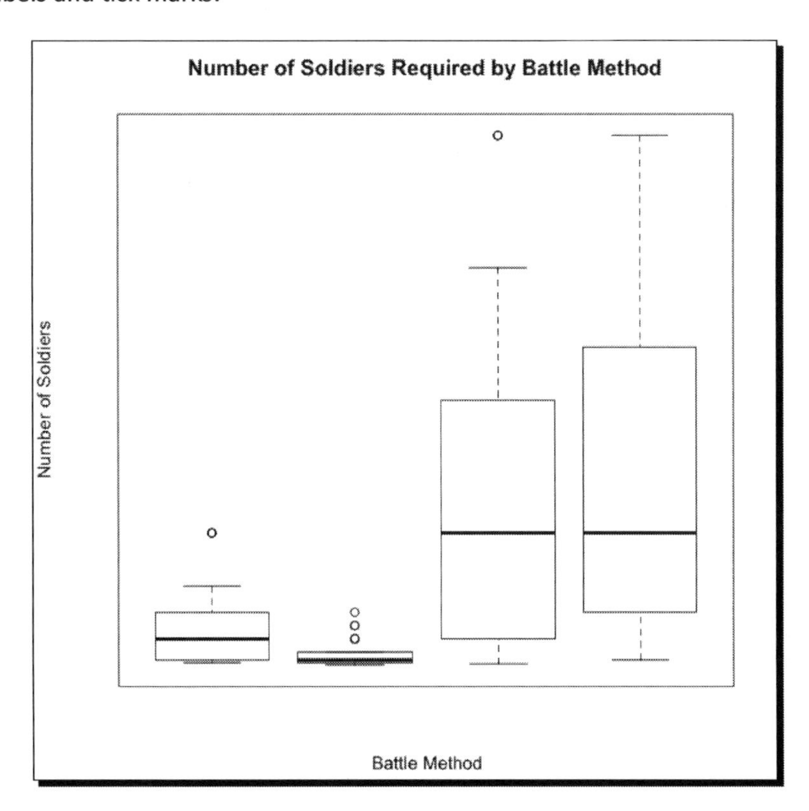

3. Use `axis(...)` to create custom axes for the box plot:

```
> #use axis(...) to add custom x and y axes to the box plot
> #your custom axes will be drawn atop the plot that is displayed
in the graphic window
> #your axes will be displayed when the axis(...) function is
executed in the R console
> #custom x axis
> axis(side = 1, at = c(1, 2, 3, 4), labels = c("Ambush",
"Fire", "Head to Head", "Surround"), las = 0)
> #custom y axis
> axis(side = 2, at = c(1000, 25000, 50000, 75000, 100000),
las = 0)
```

Your custom axes will be added to the existing plot:

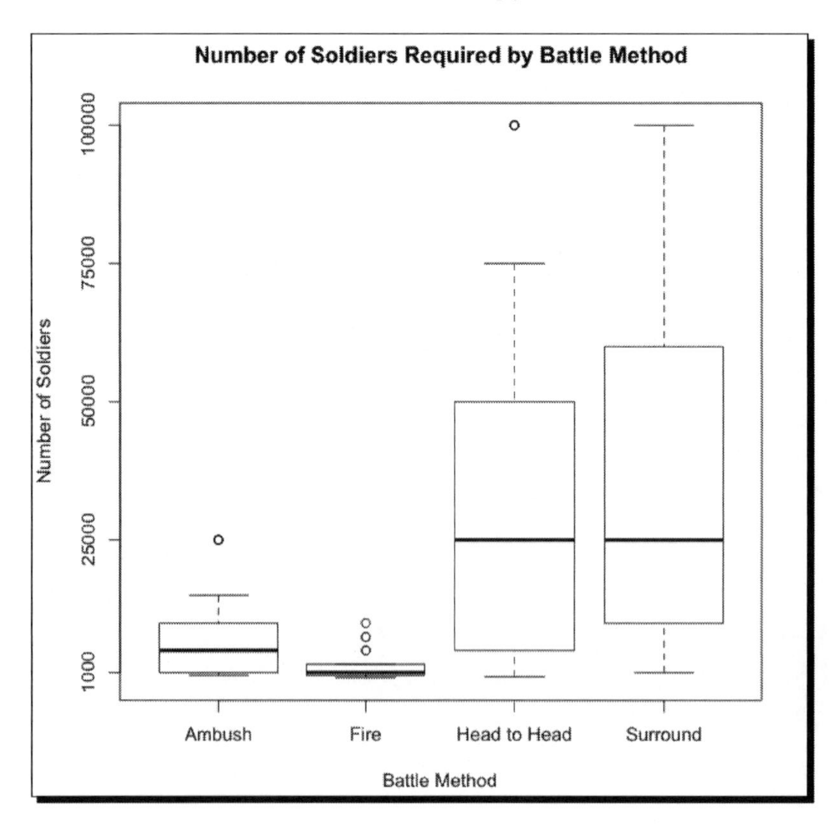

What just happened?

We customized our box plot to make it more presentable. Let us review the customization options offered by the `boxplot(...)` function.

range

We used the `boxplot(...)` function's `range` argument to alter the length of each box's whiskers. In general, `range` will take on a positive value between 0 and 1.5. At 0, a box's whiskers will extend all the way to the most extreme data points. At a value greater than 0, the boxes' whiskers will reach data points within one **interquartile range** times the `range` value. An interquartile range is the distance between the top (third quartile) and bottom (first quartile) of a given box. This measure gives us an idea of how spread out the data are. By using a range value closer to 0, we are shortening our boxes' whiskers and excluding more extreme data points. On the other hand, a higher `range` value will include more extreme points and lengthen each box's whiskers. In our case, we used a `range` value of 1:

```
> boxPlotAllMethodsShuSoldiersRange <- 1
```

This shortened our whiskers by extending them to points no more than one interquartile range beyond their boxes. Note that we also revised our `ylim` argument to better display our boxes, given the new `range`:

```
> boxplot(formula = boxplotAllMethodsShuSoldiersData,
main = boxPlotAllMethodsShuSoldiersLabelMain,
xlab = boxPlotAllMethodsShuSoldiersLabelX,
ylab = boxPlotAllMethodsShuSoldiersLabelY,
ylim = boxPlotAllMethodsShuSoldiersLimY,
range = boxPlotAllMethodsShuSoldiersRange)
```

axis(...)

To further improve our plot's aesthetics, we revised its x-axis and y-axis labels. Before adding our own axes, we had to eliminate the default ones generated by R. This entailed giving the `xaxt` and `yaxt` arguments an `n` value:

```
> boxplotAllMethodsShuSoldiersAxtX = "n"
> boxplotAllMethodsShuSoldiersAxtY = "n"
```

Subsequently, we redrew our box plot without x-axis and y-axis:

```
> boxplot(formula = boxplotAllMethodsShuSoldiersData,
main = boxPlotAllMethodsShuSoldiersLabelMain,
xlab = boxPlotAllMethodsShuSoldiersLabelX,
ylab = boxPlotAllMethodsShuSoldiersLabelY,
ylim = boxPlotAllMethodsShuSoldiersLimY,
```

```
range = boxPlotAllMethodsShuSoldiersRange,
xaxt = boxplotAllMethodsShuSoldiersAxtX,
yaxt = boxplotAllMethodsShuSoldiersAxtY)
```

We then used the `axis(...)` function twice, once for the x-axis and once for y-axis, to customize our plot's axis labels. The `axis(...)` function accepts several optional arguments, a number of which were employed in the creation of our plot:

- `side` refers to the placement of the axis, where:
    - 1 = left
    - 2 = bottom
    - 3 = top
    - 4 = right
- `at` contains a vector that holds the tick mark values for the axis
- `labels` contains a vector of text items that will be paired with the `at` values; if undefined, the `at` values will be displayed on the axis
- `las` positions the `labels` either parallel (`0`) or perpendicular (`1`) to the axis; note that `las` refers to the *label style* of the axis

When executed, the `axis(...)` function draws a new axis atop the visual currently displayed in the graphic window. For instance:

```
> axis(side = 2, at = c(10, 20, 30), labels = c("a", "b", "c"),
las = 1)
```

The code would draw a new x-axis on the *bottom* of the chart with tick marks at 10, 20, and 30 paired with the *labels* a, b, and c, that have been oriented *vertically*. Similarly, we used the following code to customize our x and y axes:

```
> #custom x axis
> axis(side = 1, at = c(1, 2, 3, 4), labels = c("Ambush", "Fire",
"Head to Head", "Surround"), las = 0)
> #custom y axis
> axis(side = 2, at = c(1000, 25000, 50000, 75000, 100000),
las = 0)
```

Our custom x-axis was placed at the bottom of the plot and effectively renamed our four boxes to *Ambush*, *Fire*, *Head to head*, and *Surround*. Our custom y-axis was placed on the left side of the plot and incorporated more meaningful soldier values than were present in the default axis.

Pop quiz

1. In `boxplot(...)`, a `range` argument of `0` would have what effect?
    a. It would eliminate the whiskers.
    b. It would extend the whiskers to the most extreme data points.
    c. It would eliminate the boxes.
    d. It would extend the boxes to the most extreme data points.
2. Which of the following is not true of the `axis(...)` function?
    a. It accepts several optional arguments.
    b. It allows for the creation of axes in four different positions.
    c. It will use the `labels` argument by default when the `at` argument is undefined.
    d. It draws atop the visual that is currently displayed in the graphic window.

Have a go hero

Create a box plot that depicts the relationship between the number of Wei soldiers targeted by each of the four battle methods. Be sure to customize your plot to improve its readability and emphasize its most important features.

Time for action – customizing a histogram

In this section, we will practice customizing the bars of a histogram and create an alternative style of histogram:

1. Use the `breaks` argument to separate the histogram's columns along the x-axis:

```
> #modify the chapter 8 histogram that depicted the frequency
distribution of past fire attack durations
> #use the breaks argument to divide the histogram's columns along
the x axis
> #breaks accepts a vector containing the points at which columns
should occur
> histFireDurationBreaks <- c(0:14)
> #use hist(...) to create and display the histogram
> hist(x = histFireDurationData,
main = histFireDurationDataMain,
xlab = histFireDurationLabX,
col = histFireDurationRainbowColor,
breaks = histFireDurationBreaks)
```

Your histogram will be displayed in the graphic window, as shown in the following:

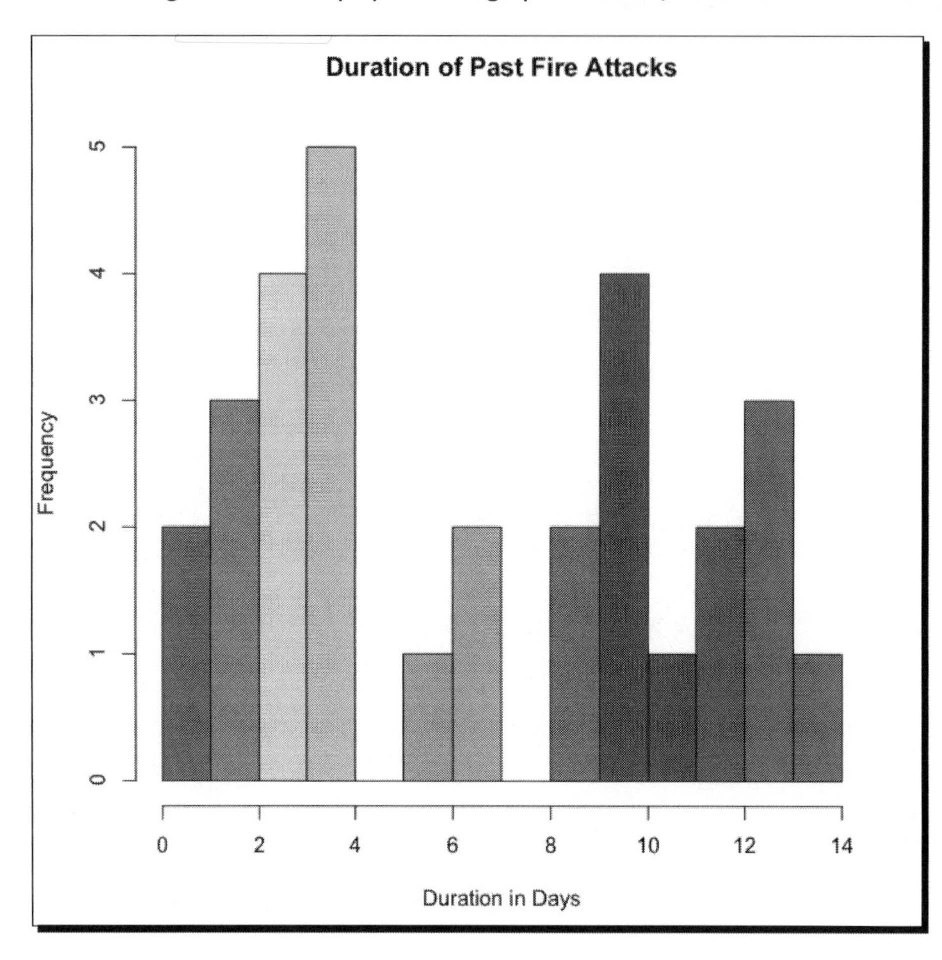

2. Use the `freq` argument to plot densities instead of counts:

```
> #use the freq argument to plot densities or counts
> #if freq is TRUE (default), counts are graphed on the y axis
> #a count tells us the number of times that a data point occurred
> #if freq is FALSE, densities are graphed on the y axis
> #a density tells us what percentage a data point's count
represents out of all occurrences
> #when summed, the densities always add up to 1
> histFireDurationFreq <- FALSE
> #remember to modify the ylim argument, as our previous one
applied to counts and not to densities
> histFireDurationDensityLimY <- c(0, 0.2)
```

```
> #use hist(...) to create and display the histogram
> hist(x = histFireDurationData,
main = histFireDurationDataMain,
xlab = histFireDurationLabX,
ylim = histFireDurationDensityLimY,
col = histFireDurationRainbowColor,
breaks = histFireDurationBreaks,
freq = histFireDurationFreq)
```

Your histogram will be displayed in the graphic window, as shown in the following:

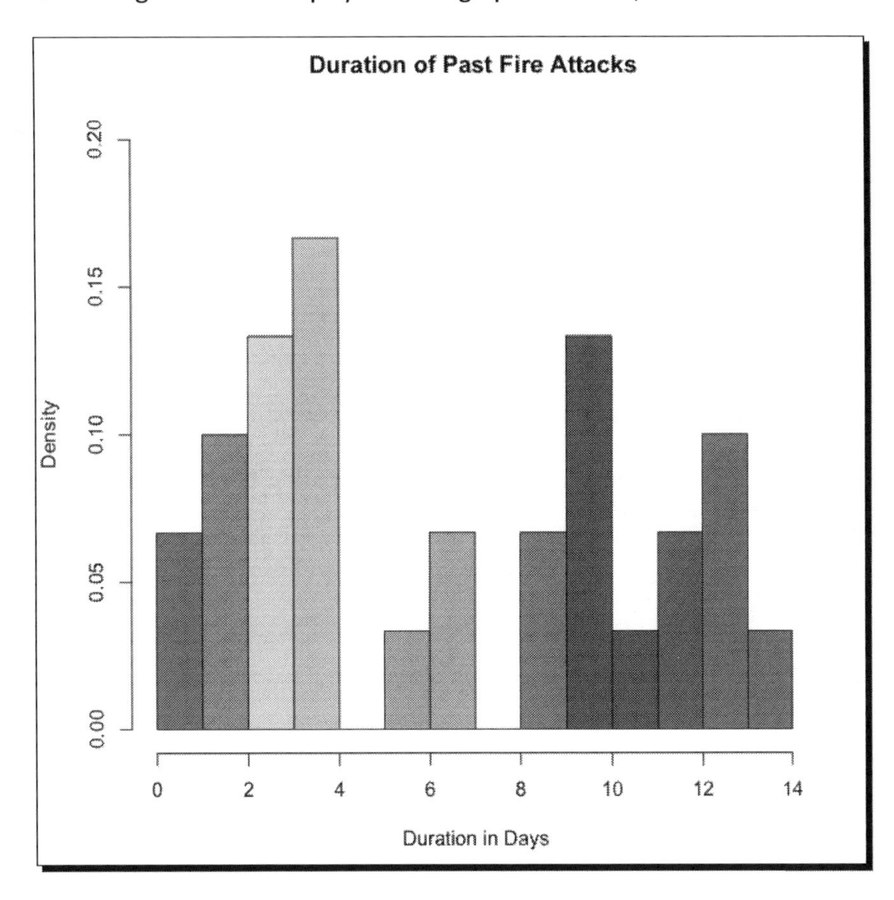

What just happened?

We set the `breaks` argument to add detail to our histogram, then defined the `freq` argument to change the display of our graphic. Let us discuss each of these actions.

breaks

The `breaks` argument is used to define where a histogram's columns are separated along the x-axis. This argument receives a vector containing the points at which the column divisions should occur. Within the `hist(...)` function, employing the breaks argument may resemble using the `xlim` argument in other graphics. However, while `xlim` rescales the x-axis of a histogram, it does not modify its columns. Therefore, the `breaks` argument is necessary when we want to define the exact points at which our columns should occur.

By default, R provided us with seven bars that spanned a width of two days each. With number-colon-number notation (`0:14`) and the `breaks` argument, we created 14 columns that spanned 1 day each:

```
histFireDurationBreaks <- c(0:14)
```

This had the effect of increasing the interpretability and detail of our histogram:

```
hist(x = histFireDurationData, main = histFireDurationDataMain,
xlab = histFireDurationLabX, col = histFireDurationRainbowColor,
breaks = histFireDurationBreaks)
```

freq

The `freq` argument allows us to toggle our histogram between displaying counts (or frequencies) and densities (or percentages). A **count** indicates how many times a value occurs within a dataset. A **density** indicates the percentage that the count of a value makes up in the entire dataset.

For instance, in the vector `c(1, 1, 1, 3, 5)`, the number 1 has a count of 3 because it occurs 3 times. The number 1 has a a density of 0.6 (or 60%) because its count of 3 makes up 3/5 of the overall dataset.

By default, `freq` is set to `TRUE` and displays counts. If it is set to `FALSE`, then densities will be graphed instead. The sum of the densities in a histogram will always equal 1, which represents 100% of the dataset.

We modified our original histogram to display densities by setting the the `freq` argument to `FALSE`:

```
histFireDurationFreq <- FALSE
```

Note that we also adjusted our `ylim` argument to appropriately display our density values:

```
histFireDurationDensityLimY <- c(0, 0.2)
```

These alterations allowed us to visualize our battle durations as percentages rather than counts:

```
hist(x = histFireDurationData,
main = histFireDurationDataMain,
xlab = histFireDurationLabX,
ylim = histFireDurationDensityLimY,
col = histFireDurationRainbowColor,
breaks = histFireDurationBreaks,
freq = histFireDurationFreq)
```

Pop quiz

1. When using `hist(...)`, what is the relationship between the `xlim` and `breaks` arguments?
   a. `breaks` sets the overall scale of the x-axis, whereas `xlim` divides the histogram's columns along the x-axis.
   b. `xlim` sets the overall scale of the x-axis, whereas `breaks` divides the histogram's columns along the x-axis
   c. `breaks` replaces the `xlim` argument when creating a histogram.
   d. `xlim` replaces the `breaks` argument when creating a histogram.
2. What is the relationship between a count and a density value?
   a. A count is the number of times that a value occurs in a dataset, whereas a density is the total count of all values in a dataset.
   b. A density is the number of times that a value occurs in a dataset, whereas a count is the total count of all values in a dataset.
   c. A count is the number of times that a value occurs in a dataset, whereas a density is the percentage of the dataset that a value accounts for.
   d. A density is the number of times that a value occurs in a dataset, whereas a count is the percentage of the dataset that a value accounts for.

Have a go hero

Create a histogram that conveys the number of Shu soldiers engaged in past fire attacks. Improve its readability by incorporating the `breaks` argument into your `hist(...)` function. Then, create a density version of the histogram using the `freq` argument. Compare your frequency and density histograms. Which do you feel is better for displaying this particular data?

Time for action – customizing a pie chart

Moving on to pie charts, we will learn how to add custom label text to a pie's slices:

1. Use the `labels` argument to add percentages to the pie chart.

```
> #modify the chapter 8 pie chart that depicted the gold cost of
the fire attack in relation to the total funds allotted to the Shu
army
> #use the labels argument to add percentages to a pie chart
> #create a vector containing the labels to be used for the pie's
slices
> pieFireGoldCostLabelsPercent <- round(pieFireGoldCostSlices /
sum(pieFireGoldCostSlices) * 100, 1)
> #use the paste(...) function to add a percent sign (%) to the
end of each label
> pieFireGoldCostLabelsPercent <-
paste(pieFireGoldCostLabelsPercent, "%", sep="")
> #note that paste(...) can be used to add any kind of text before
or after a label
> #use the pie(...) function to create and display the pie chart
> pie(x = pieFireGoldCostSlices,
labels = pieFireGoldCostLabelsPercent,
main = pieFireGoldCostMain,
col = pieFireGoldCostSpecificColors)
```

Your chart will be displayed in the graphic window, as follows:

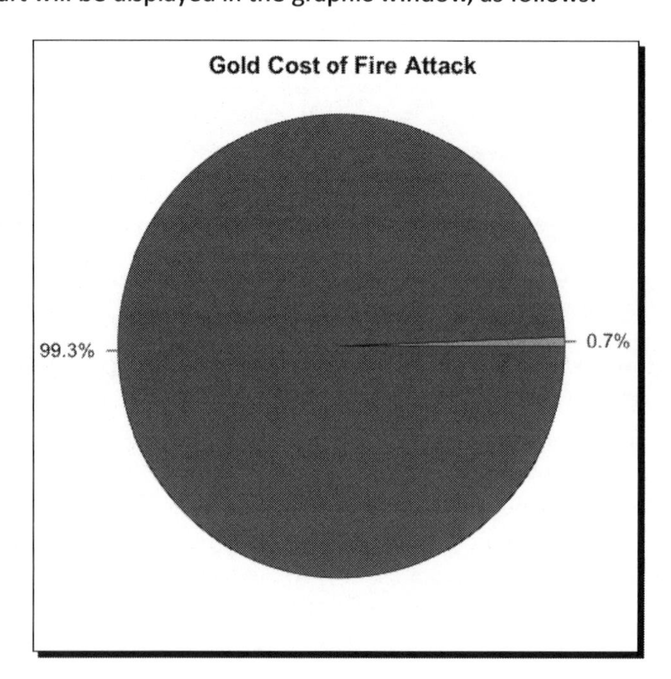

2. Add a legend to the chart:

```
> #add a legend to the pie chart
> legend(x = "bottom", legend = pieFireGoldCostLabels,
fill = pieFireGoldCostSpecificColors)
```

Your legend will be added to the existing chart, the final pie chart should look like the following:

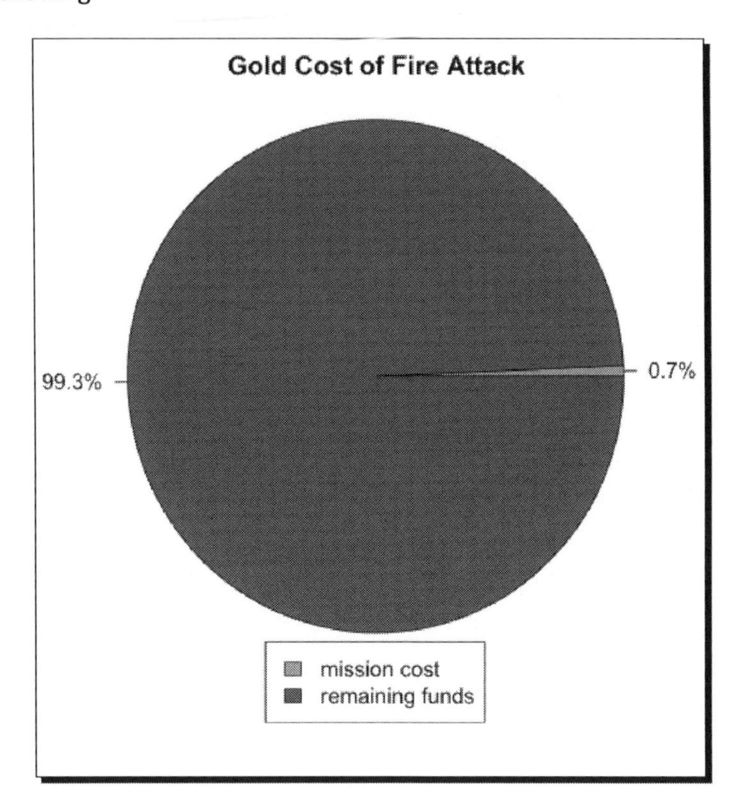

What just happened?

We just customized our pie chart by taking advantage of a new labeling option. Let us discuss how this feature is implemented.

Custom labels

We revised our pie chart's labels to display percentage values, rather than raw gold amounts. To accomplish this, we calculated the necessary percentages using the `round(x, digits)` function in tandem with some routine mathematics. In the `round(x, digits)` function `x` is a number, and `digits` is the number of decimal places that `x` should be rounded to.

Therefore, rounding the number 1.2345 using:

```
> round(1.2345, 2)
```

The code would yield an output of:

```
[1] 1.23
```

For our chart, `x` contained a formula that yielded the percentage that each slice represents out of our total. The `digits` argument dictated that this percentage be rounded to a single decimal point:

```
> pieFireGoldCostLabelsPercent <- round(pieFireGoldCostSlices /
sum(pieFireGoldCostSlices) * 100, 1)
```

To improve the readability of our percentages, we then used the `paste(...)` function to append a percent sign (`%`) to each of our labels. In our activity, the `paste(...)` function included the following arguments:

- `originalValues`: a vector containing the items that are to be appended
- `appendText`: the text to be added to the original values
- `sep`: an optional separator between the original value and the appended text; a single space by default

Hence, the general `paste(...)` function takes on the following form:

```
paste(originalValues, appendText, sep = "sep")
```

Thus, if we were to enter the following code into the R console:

```
> paste(c("a", "b"), "c", sep = "/")
```

Our resulting output would be:

```
[1] "a/c" "b/c"
```

We used the `paste(...)` function to append a percentage sign (`%`) to each of our percentage labels (`pieFireGoldCostLabelsPercent`) and indicated that they should not be separated by any blank space or characters (`sep = ""`):

```
> pieFireGoldCostLabelsPercent <- paste(pieFireGoldCostLabelsPercent,
"%", sep="")
```

Lastly, our `pie(...)` function incorporated our custom percentage labels:

```
> pie(x = pieFireGoldCostSlices,
labels = pieFireGoldCostLabelsPercent,
main = pieFireGoldCostMain,
col = pieFireGoldCostSpecificColors)
```

Note that the `paste(...)` function can be used to add any kind of text to a label. Its general purpose is to append text to the front and back of values. As such, it is applicable in many situations.

legend(...)

Yet again, we found it necessary to include a legend in our chart. Without a legend, our graphic would not indicate what our percentage labels referred to. Our legend was placed at the bottom of our graphic and reflected our chart's original text labels (rather than percentages) and colors; the following is the code:

```
> legend(x = "bottom", legend = pieFireGoldCostLabels,
fill = pieFireGoldCostSpecificColors)
```

Pop quiz

1. What would be the result of the following `round(x, digits)` function?

   ```
   > round(9.876543, 3)
   ```

   a. 9.877

   b. 9.876

   c. 9.87

   d. 9.88

2. In the `paste(originalValues, appendText, sep = "sep")` function, what does the `sep` argument represent?

   a. A vector containing the items that are to be appended.

   b. The text to be added to the original values.

   c. An optional separator between the original value and the appended text.

   d. A vector containing the text to be appended.

Have a go hero

Create a pie chart that conveys the relationship between the number of soldiers engaged in the planned fire attack and the total number of soldiers housed at Hanzhong. Be sure to experiment with the customization options that we have covered in our previous examples.

Time for action – building a graphic

Having explored an extensive range of graphic types and customizations in R, our next challenge is to build a graphic from the ground up. To accomplish this feat, we will start with an empty foundation and use our customization arguments and functions to build a complete graphic:

1. Use the `plot(...)` function to create a foundation for the graphic:

```
> #build a custom graphic from scratch
> #step 1: create a foundation
> #create a graphic that depicts the number of Shu and Wei
soldiers engaged in past fire attacks
> #prepare the graphic's basic parameters
> #note that this will require thinking ahead about the
information that you want to display
> buildFireSoldiersMain <- "Soldiers Engaged by Kingdom"
> buildFireSoldiersLabX <- "Battle Number"
> buildFireSoldiersLabY <- ""
> buildFireSoldiersLimX <- c(0, 30)
> buildFireSoldiersLimY <- c(0, 50000)
> #hide the points and axes
> buildFireSoldiersType <- "n"
> buildFireSoldiersAxtX <- "n"
> buildFireSoldiersAxtY <- "n"
> #use the plot(...) function to create a foundation for the
graphic
> plot(x = 0, y = 0, main = buildFireSoldiersMain,
xlab = buildFireSoldiersLabX, ylab = buildFireSoldiersLabY,
xlim = buildFireSoldiersLimX, ylim = buildFireSoldiersLimY,
type = buildFireSoldiersType, xaxt = buildFireSoldiersAxtX,
yaxt = buildFireSoldiersAxtY)
```

An empty foundation for our graphic will open in the graphic window, as shown:

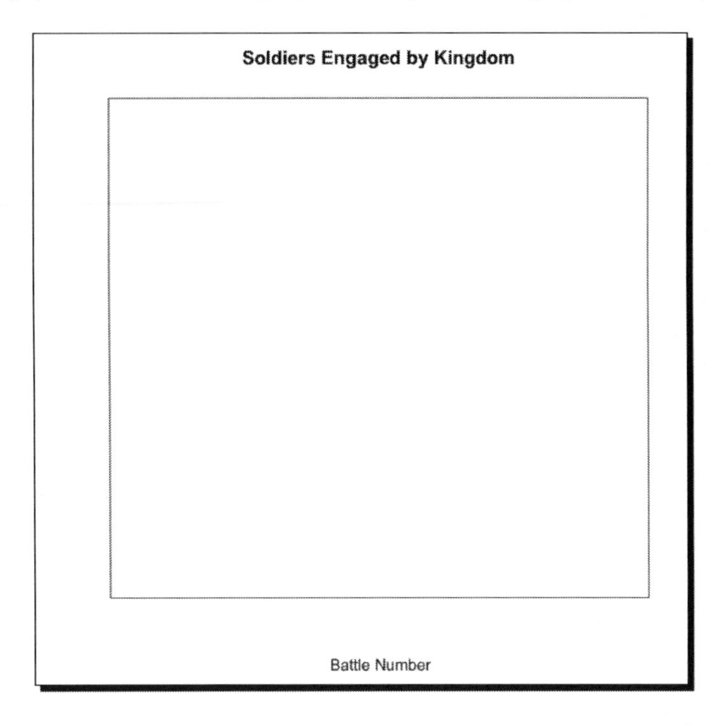

2. Add axes to the graphic.

```
> #step 2: add axes
> #use axis(...) to add custom x and y axes to the graphic
> #x axis
> axis(side = 1, at = c(0:30), las = 0)
> #y axis
> axis(side = 2,
at = c(1000, 5000, 10000, 20000, 30000, 40000, 50000),
las = 1)
```

Your custom axes will be added to the existing graphic, and will look like the following:

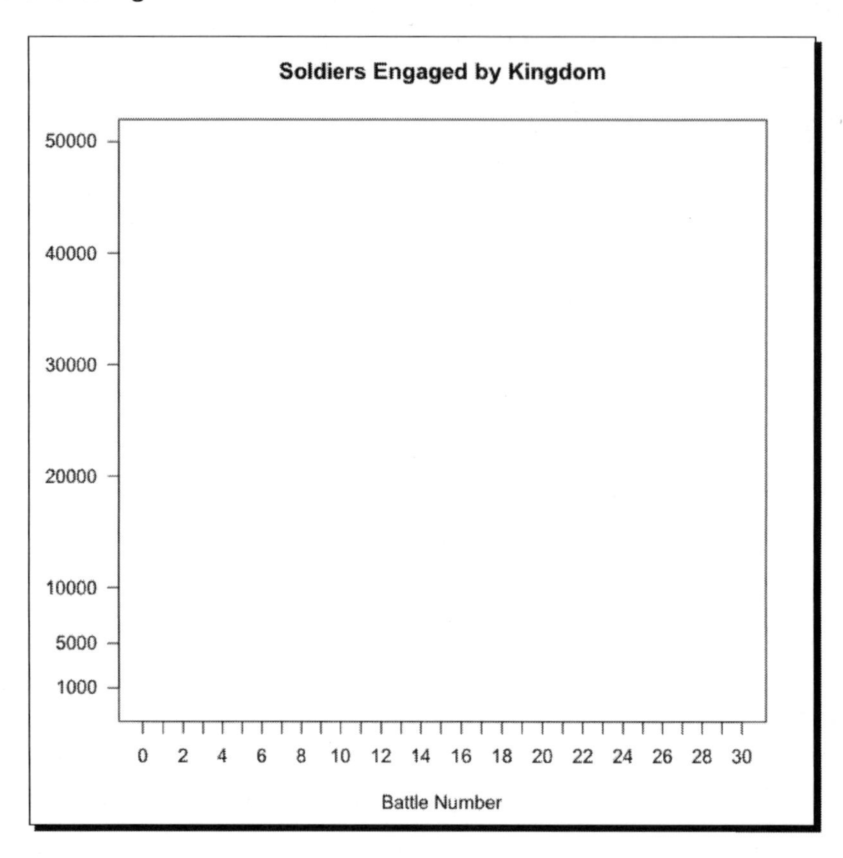

3. Add data to the graphic:

```
> #step 3: add data
> #use points(...) to add data to the graphic
> #note that lines(...) can also be used to add data to a graphic
> #add points to show the number of Shu soldiers engaged in past
fire attacks
> pointsFireShuSoldiersDataX <- c(1:30)
> pointsFireShuSoldiersDataY <- subsetAmbush$ShuSoldiersEngaged
> pointsFireShuSoldiersType <- "p"
> pointsFireShuSoldiersColor <- "forestgreen"
> points(x = pointsFireShuSoldiersDataX,
y = pointsFireShuSoldiersDataY,
type = pointsFireShuSoldiersType,
col = pointsFireShuSoldiersColor)
```

```
> #add points to show the number of Wei soldiers engaged in past
fire attacks
> pointsFireWeiSoldiersDataX <- c(1:30)
> pointsFireWeiSoldiersDataY <- subsetAmbush$WeiSoldiersEngaged
> pointsFireWeiSoldiersType <- "p"
> pointsFireWeiSoldiersColor <- "blue"
> pointsFireWeiSoldiersPch <- 0
> points(x = pointsFireWeiSoldiersDataX,
y = pointsFireWeiSoldiersDataY,
type = pointsFireWeiSoldiersType,
col = pointsFireWeiSoldiersColor,
pch = pointsFireWeiSoldiersPch)
```

Your custom points will be added to the graphic, as shown in the following:

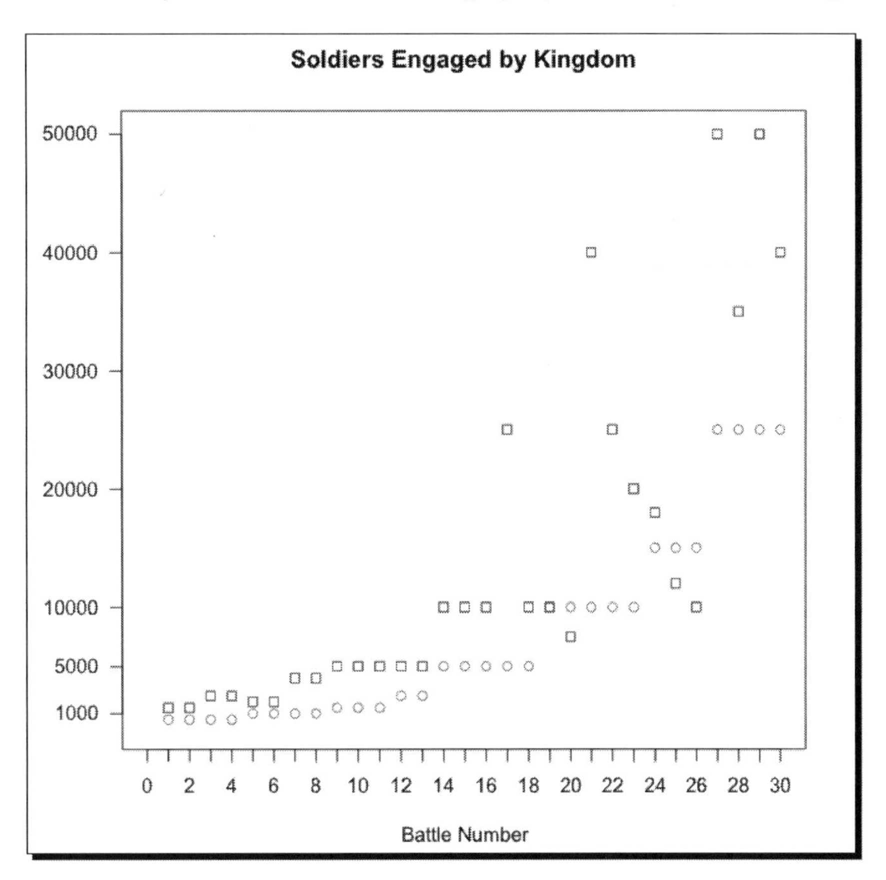

4. Add a legend to the graphic:

```
> #step 4: add a legend, if necessary
> #use legend(...) to add a legend to the graphic
> legend(x = 0, y = 50000, legend = c("Shu", "Wei"),
fill = c(pointsFireShuSoldiersColor,
pointsFireWeiSoldiersColor))
```

Your legend will be added to the graphic. The final graphic will look like the following:

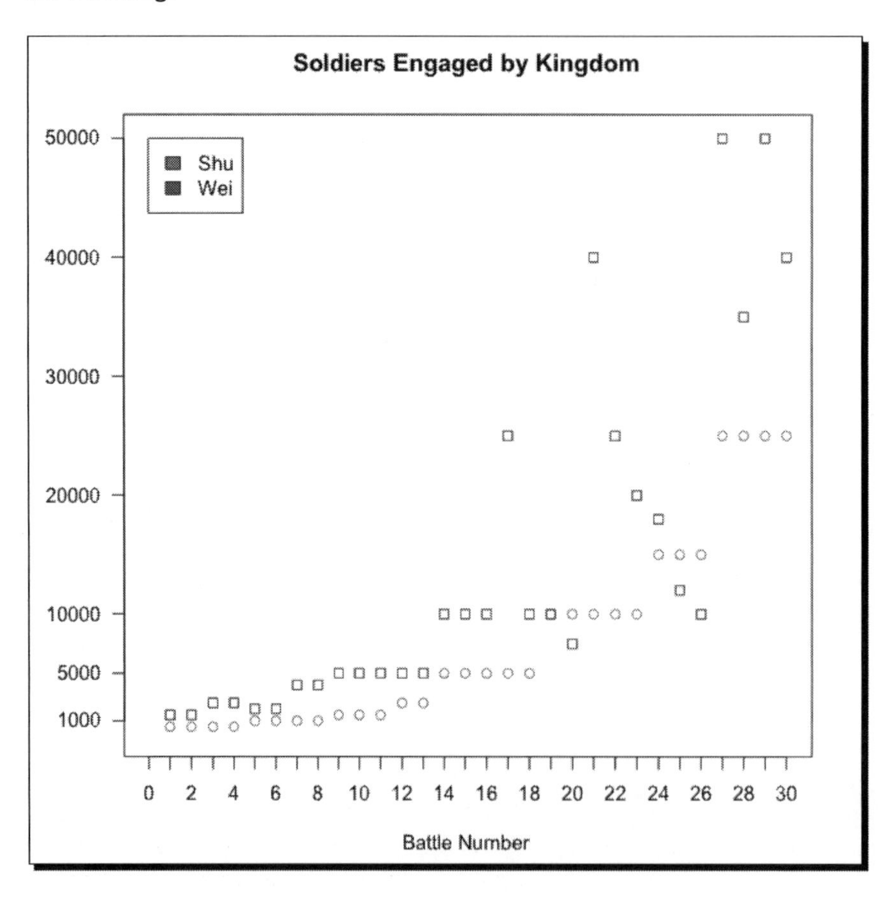

What just happened?

We used our custom graphics functions to build an entire graphic from scratch. Let us review the steps involved in this process:

1. **Building the foundation**

   We began by using our `plot(...)` function to create a foundation for our new graphic. The main difference when creating a foundation graphic compared to a normal one is that we do not want to display any data. Hence, we set the `x` and `y` values to `0`, the type to `n`, and the `xaxt` and `yaxt` arguments to `n`. This yields a blank square to which we can add custom information later. However, it is still critical to match the `xlim` and `ylim` arguments to the bounds of the data that we plan to use, in spite of the fact that the axes themselves are hidden. The overall title and x and y labels can be optionally defined, if we would like them to appear on the graphic:

```
> buildFireSoldiersMain <- "Soldiers Engaged by Kingdom"
> buildFireSoldiersLabX <- "Battle Number"
> buildFireSoldiersLabY <- ""
> buildFireSoldiersLimX <- c(0, 30)
> buildFireSoldiersLimY <- c(0, 50000)
> buildFireSoldiersType <- "n"
> buildFireSoldiersAxtX <- "n"
> buildFireSoldiersAxtY <- "n"
```

> Note that this step required that we think ahead about the data that we wanted to display, especially as it pertains to the x-axis and y-axis limits. At times, this preparation step may also call for experimentation with default graphics to lend us a better idea of how the data will display when customized.

   Our `plot(...)` function incorporated all of these parameters and rendered us with a foundation graphic that was prepared to incorporate our data:

```
> plot(x = 0, y = 0, main = buildFireSoldiersMain,
xlab = buildFireSoldiersLabX, ylab = buildFireSoldiersLabY,
xlim = buildFireSoldiersLimX, ylim = buildFireSoldiersLimY,
type = buildFireSoldiersType, xaxt = buildFireSoldiersAxtX,
yaxt = buildFireSoldiersAxtY)
```

2. **Adding axes**

   Since we completely hid our axes from the foundation graphic, our next step involves creating custom ones. We added custom axes to our graphic using the familiar `axis(...)` function:

```
> #x axis
> axis(side = 1, at = c(0:30), las = 0)
> #y axis
> axis(side = 2,
at = c(1000, 5000, 10000, 20000, 30000, 40000, 50000),
las = 1)
```

   For a review of the `axis(...)` function, see the *Customizing a box plot* section of this chapter.

3. **Adding data**

   After the axes are defined, it is time to add data to the graphic. While we chose to use the `points(...)` function to add data in this activity, note that `lines(...)` could also be used, depending on the type of visual effect that you prefer. Our data consisted of the number of Shu and Wei soldiers engaged in past fire attacks:

```
> #add points to show the number of Shu soldiers engaged in past
fire attacks
> pointsFireShuSoldiersDataX <- c(1:30)
> pointsFireShuSoldiersDataY <- subsetAmbush$ShuSoldiersEngaged
> pointsFireShuSoldiersType <- "p"
> pointsFireShuSoldiersColor <- "forestgreen"
> points(x = pointsFireShuSoldiersDataX, y =
pointsFireShuSoldiersDataY, type = pointsFireShuSoldiersType,
col = pointsFireShuSoldiersColor)
> #add points to show the number of Wei soldiers engaged in past
fire attacks
> pointsFireWeiSoldiersDataX <- c(1:30)
> pointsFireWeiSoldiersDataY <- subsetAmbush$WeiSoldiersEngaged
> pointsFireWeiSoldiersType <- "p"
> pointsFireWeiSoldiersColor <- "blue"
> pointsFireWeiSoldiersPch <- 0
> points(x = pointsFireWeiSoldiersDataX,
y = pointsFireWeiSoldiersDataY,
type = pointsFireWeiSoldiersType,
col = pointsFireWeiSoldiersColor,
pch = pointsFireWeiSoldiersPch)
```

   For a review of `points(...)` and `lines(...)` functions, see the *Customizing a scatterplot* and *Customizing a line chart* sections of this chapter.

4. **Add a legend, if necessary**

   The final step in building a custom graphic is to add a legend, if the graphic that you have created calls for one. This can be done using the same `legend(...)` function that we have exercised throughout our time exploring R's graphic capabilities:

   ```
   > legend(x = 0, y = 50000, legend = c("Shu", "Wei"),
   fill = c(pointsFireShuSoldiersColor,
   pointsFireWeiSoldiersColor))
   ```

   For a review of the `legend(...)` function, see the *Customizing graphics* section of the previous chapter.

In the end, we managed to build a complete, fully customized scatterplot starting from scratch. This development process is invaluable when you are creating graphics to present to others, use in reports, or publish outside of R.

Pop quiz

1. Is it important to define the `xlim` and `ylim` arguments on a foundation graphic? Why or why not?
   a. It is important, because these arguments scale our axes in preparation for data that will be added later.
   b. It is important, because the `plot(...)` function will not execute if these arguments are left undefined.
   c. It is not important, because the default axes will automatically scale to our data.
   d. It is not important, because the x and y axes are hidden.
2. Could steps 2 (adding axes), 3 (adding data), and 4 (adding legend) of the graphic building process occur in a different order than the one that was demonstrated in this section?
   a. No, they must be executed precisely in the order specified.
   b. Yes, axes or data can occur in any order, but a legend must be added last.
   c. Yes, the data or legend can occur in any order, but axes must be added first.
   d. Yes, these steps merely add visual elements to our graphic and therefore can be executed in any order.

Have a go hero

You have practiced generating highly customized graphics and even learned to build your own graphic from scratch. Use your refined R talents to create at least three graphics that will convince the top generals of the Shu army to join you in battle. Recall that the generals are most interested in scrutinizing the details of your proposed attack and comparing it with alternative combat strategies. Be sure to explore new combinations of graphic arguments and functions. Refer back to the individual sections of this chapter for assistance with creating graphics of particular types.

Time for action – building a graphic with multiple visuals

Within R, it is possible to generate graphics that are composed from two or more separate visuals. Let us build a graphic that displays several pieces of information about our fire attack strategy simultaneously:

1. Prepare the graphic window to display multiple graphics simultaneously:

```
> #use par(mfcol) to prepare the graphic window to display
multiple graphics simultaneously
> #the mfcol argument receives a vector indicating the number of
rows and columns to reserve for separate graphics in the graphics
window
> #here, we want 4 total graphics, so use a 2x2 vector
> par(mfcol = c(2,2))
> #note that a blank graphic window will open
> #if this window is closed, your graphic window will default back
to displaying a single visual
> #if it remains open, your graphic window will continue to add
visuals to the 2x2 grid as they are created
```

2. Create the first graphic:

```
> #create the first graphic by duplicating the steps taken in the
Building a graphic activity
> #this scatterplot depicted the number of Shu and Wei soldiers
engaged in past fire attacks
> plot(x = 0, y = 0, main = buildFireSoldiersMain,
xlab = buildFireSoldiersLabX, ylab = buildFireSoldiersLabY,
xlim = buildFireSoldiersLimX, ylim = buildFireSoldiersLimY,
type = buildFireSoldiersType, xaxt = buildFireSoldiersAxtX,
yaxt = buildFireSoldiersAxtY)
```

```
> axis(side = 1, at = c(0:30), las = 0)
> axis(side = 2,
at = c(1000, 5000, 10000, 20000, 30000, 40000, 50000),
las = 1)

> points(x = pointsFireShuSoldiersDataX,
y = pointsFireShuSoldiersDataY,
type = pointsFireShuSoldiersType,
col = pointsFireShuSoldiersColor)
> points(x = pointsFireWeiSoldiersDataX,
y = pointsFireWeiSoldiersDataY,
type = pointsFireWeiSoldiersType,
col = pointsFireWeiSoldiersColor,
pch = pointsFireWeiSoldiersPch)

> legend(x = 0, y = 50000, legend = c("Shu", "Wei"),
fill = c(pointsFireShuSoldiersColor,
pointsFireWeiSoldiersColor))
```

Your graphic will now have additional space surrounding it, which can be used to incorporate new graphics, as shown in the following:

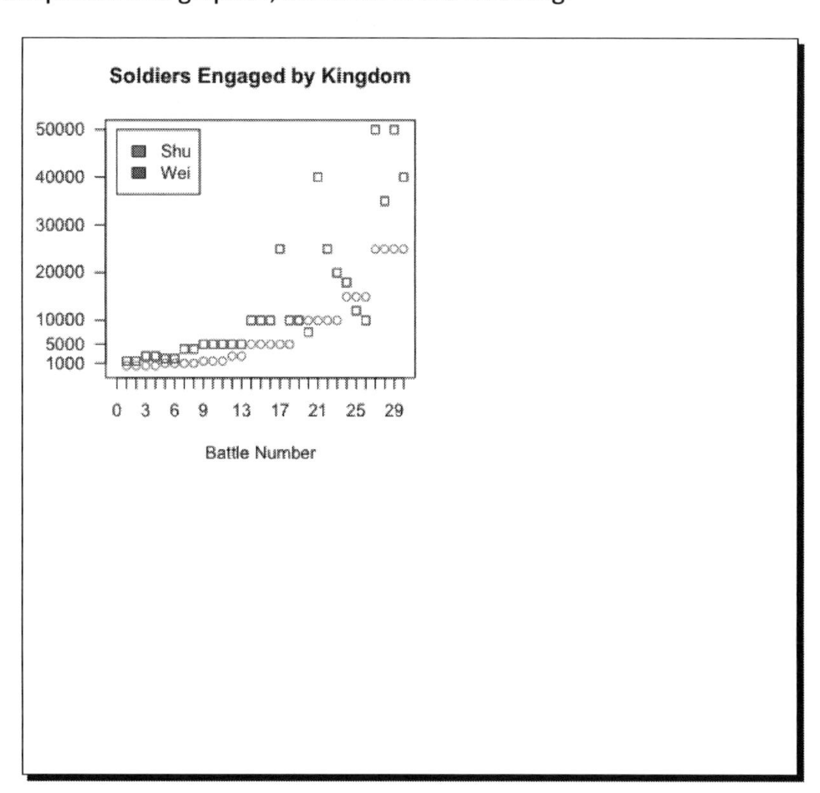

3. Add a second chart to the graphic:

```
> #add a second chart that depicts the duration of past battles
> #create new variables where necessary
> #otherwise reuse the variables from our initial graphic
> #basic parameters
> buildFireDurationMain <- "Duration in Days"
> buildFireDurationLabY <- "Days"
> buildFireDurationLimY <- c(0, 14)
> #use the plot(...) function to create a foundation for the
graphic
> plot(x = 0, y = 0, main = buildFireDurationMain,
xlab = buildFireSoldiersLabX, ylab = buildFireDurationLabY,
xlim = buildFireSoldiersLimX, ylim = buildFireDurationLimY,
type = buildFireSoldiersType, xaxt = buildFireSoldiersAxtX,
yaxt = buildFireSoldiersAxtY)

> #axes
> #x axis
> axis(side = 1, at = c(0:30), las = 0)
> #y axis
> axis(side = 2, at = c(0:14), las = 1)

> #use lines(...) to add data to the graphic
> #add a line representing the duration in days for each battle
> lineFireDurationDataX <- c(1:30)
> lineFireDurationDataY <- subsetFire$DurationInDays
> lineFireDurationType <- "o"
> lineFireDurationWidth <- 1
> lineFireDurationColor <- "red"
> lines(x = lineFireDurationDataX, y = lineFireDurationDataY,
type = lineFireDurationType, lwd = lineFireDurationWidth,
col = lineFireDurationColor)

> #use abline(...) to add a horizontal line to the chart
> #add a line representing the mean duration
> lineFireDurationMeanWidth <- 3
> lineFireDurationMeanColor <- "blue"
> abline(h = mean(lineFireDurationDataY),
lwd = lineFireDurationMeanWidth,
col = lineFireDurationMeanColor)
```

Your new chart will be added to the existing graphic, as shown in the following:

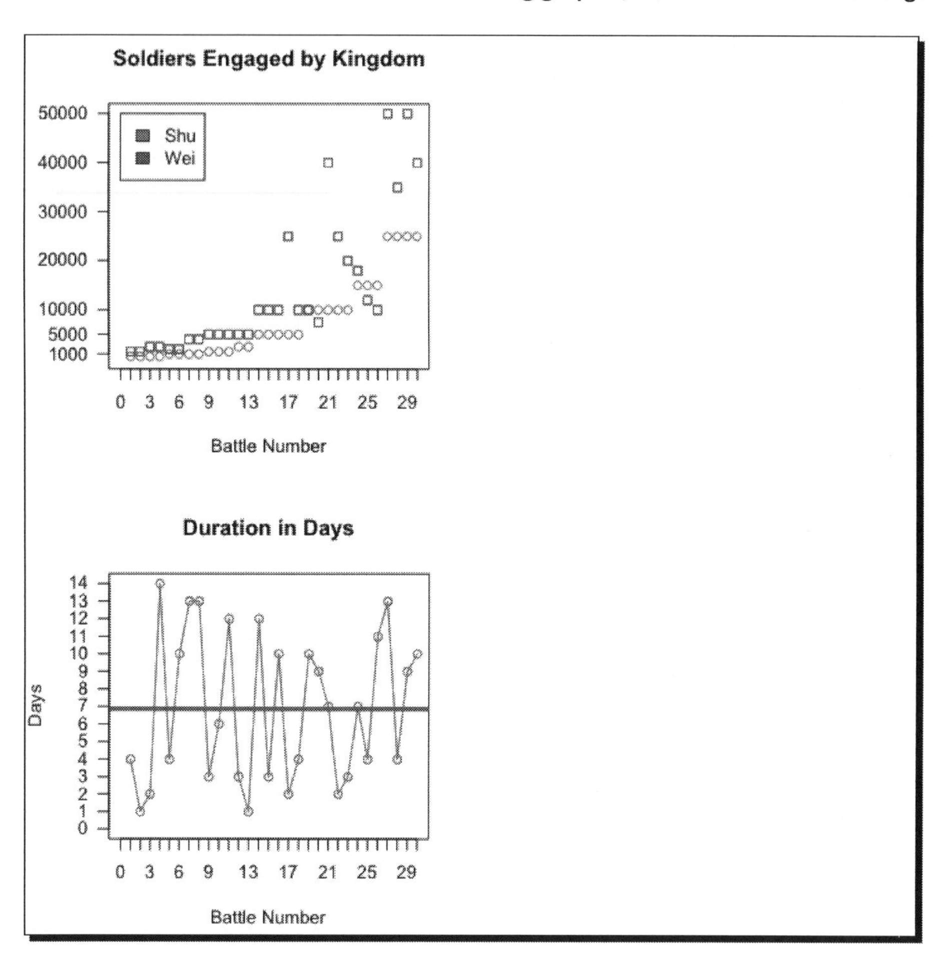

4. Add a third chart to the graphic:

```
> #add a third chart that depicts the percentage of victorious
fire attacks when the strategy is executed successfully
> #basic parameters
> buildFireResultMain <- "Result When Successfully Executed"
> buildFireResultSlices <- c(length(subset(numericResultFire,
numericExecutionFire == 1 & numericResultFire == 1)),
length(subset(numericExecutionFire,
numericExecutionFire == 1 & numericResultFire == 0)))
> buildFireResultLabels <- paste(buildFireResultSlices /
sum(buildFireResultSlices) * 100, "%", sep = "")
> buildFireResultColors <- c("red", "blue")
```

```
> #use the pie(...) function to create and display the pie chart
> pie(x = buildFireResultSlices,
labels = buildFireResultLabels,
main = buildFireResultMain,
col = buildFireResultColors)

> #legend
> legend(x = "topright", legend = c("Victory", "Defeat"),
fill = buildFireResultColors, cex = 0.85)
```

Your new chart will be added to the existing graphic, as shown in the following:

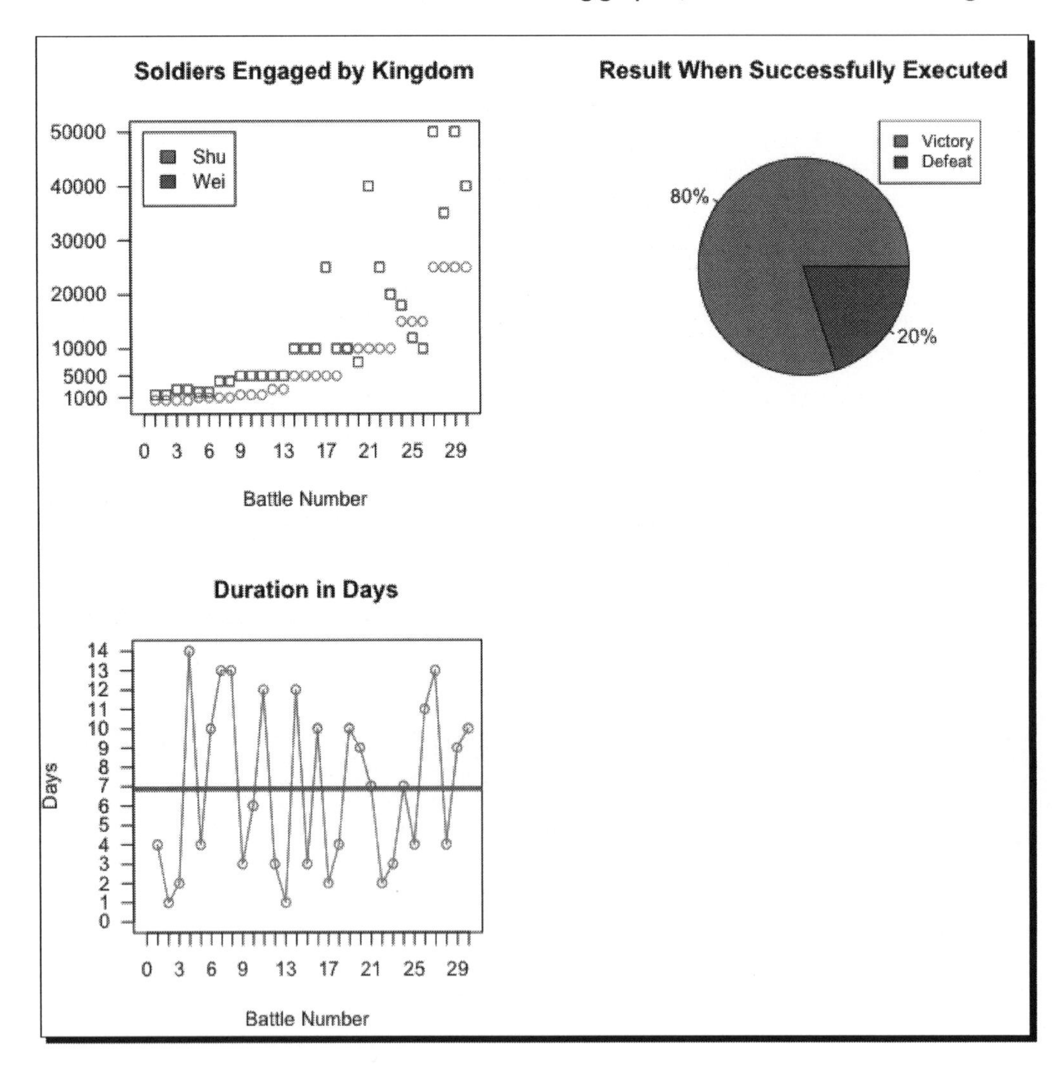

5. Add a fourth chart to the graphic:

```
> #add a fourth chart that compares the gold cost (in thousands)
of the fire attack with the other battle methods
> #get the raw cost of the various methods using comparable
resources
> goldCostFire <- functionGoldCost(2500, 225, 7)
> goldCostAmbush <- functionGoldCost(meanShuSoldiersAmbush,
225, meanDurationAmbush)
> goldCostHeadToHead <-
functionGoldCost(meanShuSoldiersHeadToHead, 225,
meanDurationHeadToHead)
> goldCostSurround <- functionGoldCost(meanShuSoldiersSurround,
225, meanDurationSurround)

> #basic parameters
> #note that the bar heights are divided by 1000 so they are
represented in thousands of gold
> #presenting larger numbers in this manner is one way to keep our
axes cleaner and our graphics more readable
> buildCostHeight <- c(goldCostFire, goldCostAmbush,
goldCostHeadToHead, goldCostSurround) / 1000
> buildCostMain <- "Cost Comparison by Method"
> buildCostLabX <- "Gold Cost (in thousands)"
> buildCostLimX <- c(0, 400)
> buildCostLimY <- c(0, 5)
> buildCostNames <- c("Fire", "Amb", "Head", "Sur")
> buildCostColors <- rainbow(length(buildCostHeight))
> buildCostHoriz <- TRUE

> #use the barplot(...) function to create and display the bar
chart
> barplot(height = buildCostHeight, main = buildCostMain,
xlab = buildCostLabX, xlim = buildCostLimX,
ylim = buildCostLimY, names = buildCostNames,
col = buildCostColors, horiz = buildCostHoriz)

> #legend
> legend(x = 275, y = 2,
legend = round(buildCostHeight * 1000, 0),
fill = buildCostColors, title = "Exact Cost", cex = 0.75)
```

Your new chart will be added to the existing graphic. The final graphic will look like the following:

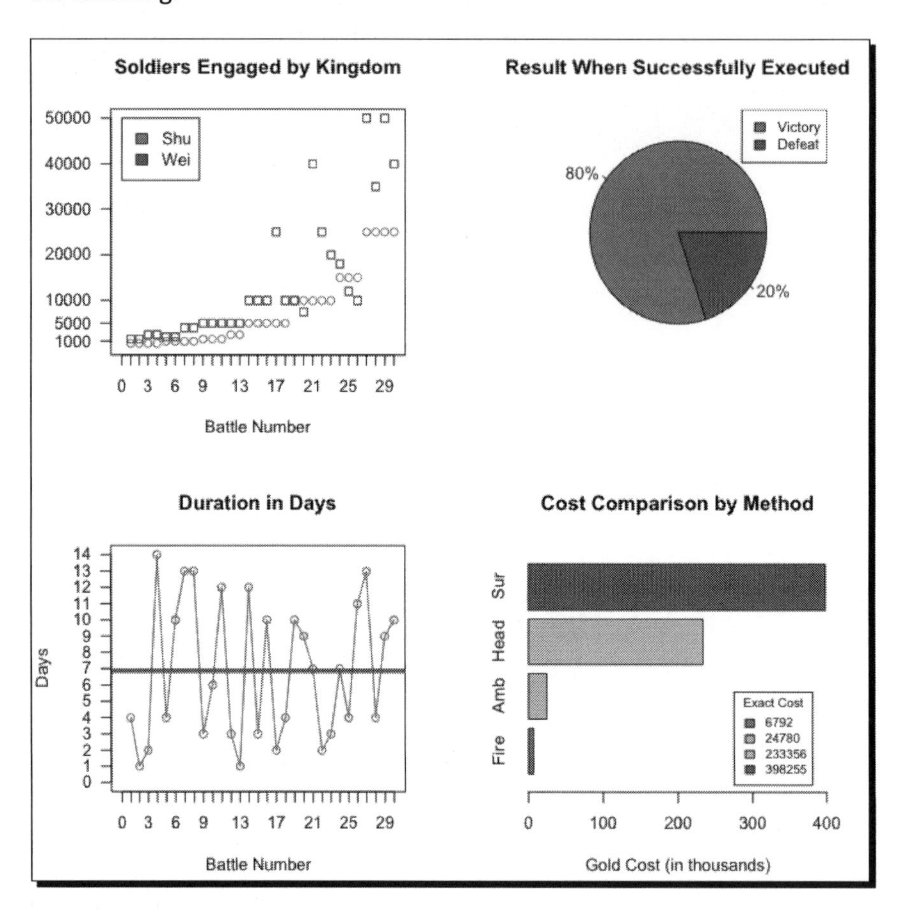

What just happened?

We built a custom visual that was composed from a set of four individual graphics.

Note that this section will only highlight the new or unique features that were encountered during this process. You should already be familiar with generating individual graphics and customizing their parameters from our previous work.

par(mfcol)

The `par(mfcol)` command modifies the number of visuals that are displayed in the graphic window. By default, the graphic window displays a single visual. The `mfcol` argument accepts a vector indicating the number of rows and columns of visuals to be displayed in the graphic window. For example:

```
> par(mfcol = c(3, 3))
```

The code would reserve space in the graphic window for nine visuals that would be displayed in a 3-row by 3-column grid. Note that the `mfcol` vector does *not* have to be symmetrical. For instance, a 5 by 1 or 2 by 10 vector would also be acceptable.

Our `par(mfcol)` command told our graphic window to display our visuals in a 2-row by 2-column grid:

```
> par(mfcol = c(2,2))
```

When `par(mfcol)` is executed, a blank graphic window will open. It is important to keep this window open. As long as it remains open, all graphics generated by R will be added to the grid defined by `par(mfcol)`. Once the graphic window is closed, it will default back to generating single visuals. At that point, `par(mfcol)` can be used again to redefine the space of the graphic window.

Graphics

Once the space of our graphic window was defined, we simply added new visuals one by one. Notice that this process is identical to creating individual graphics. The difference is that the graphic window will continue to add new visuals to the same space, rather than replacing the previous visual each time a new one is created. Thus, we are able to combine multiple visuals into a single graphic.

We started by building two charts from scratch, one depicting the soldiers engaged in fire attacks on a scatterplot, and one displaying the duration of fire attacks in a line chart. We then generated two highly customized charts, one depicting the result of fire attacks when successfully executed as a pie, and one comparing the cost of the battle methods on a bar chart. Ultimately, we arrived at a single graphic containing information from four separate visuals.

While creating the charts that composed our combined graphic, we encountered two notable items that deserve an explanation here.

Horizontal and vertical lines

The first occurred while making our fire attack duration line chart. You may have noticed that we drew a flat, horizontal line across the chart at the mean duration. To accomplish this, we used the `abline(...)` function in a new way. Previously, we used `abline(...)` to draw best fit lines on scatterplots in the *Customizing a scatterplot* section of this chapter. Here, we used the `h` argument to define a point where a horizontal line should be drawn across our chart. By setting `h` to the mean duration, we were able to visualize the average fire attack duration amidst the fluctuations experienced across each individual battle:

```
> abline(h = mean(lineFireDurationDataY),
lwd = lineFireDurationMeanWidth, col = lineFireDurationMeanColor)
```

Note that the `abline(...)` function also has a `v` argument, which can be used to define a vertical line at any point along the chart. If `h` and `v` are defined together, an intersecting pair of horizontal and vertical lines will be drawn.

Nested functions

A complex code segment that we encountered while making our pie chart involved a series of nested functions:

```
> buildFireResultSlices <- c(length(subset(numericResultFire,
numericSuccessfullyExecutedFire == 1 & numericResultFire == 1)),
length(subset(numericSuccessfullyExecutedFire,
numericSuccessfullyExecutedFire == 1 & numericResultFire == 0)))
```

Here, we created our pie's slices using a combination of the `c(...)`, `length(...)`, and `subset(...)` functions. Individually, these are all familiar. When combined, they can look confusing at first glance. The key to reading nested functions is to work from the innermost function to the outermost function. The key to creating nested functions is to remember to close your parenthesis in the opposite order that they are opened. For example, while the `c(...)` function was the first opened in our code, it was the last one that was closed. The following example illustrates this principle:

```
> function1(function2(function3()))
```

In nested functions, the innermost function is always executed first, followed by its surrounding function, and so on.

As with other programming languages, functions in R can be nested at virtually unlimited levels. On one hand, nesting makes our code more compact and efficient. On the other hand, excessive nesting makes our code unreadable and undesirably complex. Take these points into consideration when nesting functions.

Pop quiz

1. Which of the following is **not** true when of the `par(mfcol)` command?
    a. When executed, `par(mfcol)` will launch a new graphic window.
    b. Closing the graphic window will cancel the effects of the most recently executed `par(mfcol)` command.
    c. The `mfcol` argument only accepts symmetrical vectors.
    d. The `par(mfcol)` can be executed again to redefine the space of the graphic window.

2. What impact would the following line of code have on a visual displayed in the graphic window?

```
> abline(h = 4, v = 10)
```

    a. A horizontal line would be drawn at 4 on the y-axis.
    b. A vertical line would be drawn at 10 on the x-axis.
    c. A horizontal line would be drawn at 4 on the y-axis and a vertical line would be drawn at 10 on the x-axis.
    d. No lines would be drawn.

3. Which of the following code fragments demonstrates how to properly nest the `c(...)` function inside the `sum(data)` function inside the `mean(data)` function?
    a. `c(sum(data), mean(data))`
    b. `mean(sum(c(...)))`
    c. `c(sum(mean(data)))`
    d. `mean(sum(data), c(...))`

Have a go hero

Combine the visuals that you created in the previous activity into a single R graphic. Then, hold a recruitment conference with the Shu generals and convince them that your strategy is worthy of their services.

Summary

In this chapter, you created several charts, graphs, and plots to convey your battle strategy and recruit the top generals in the Shu army. To do so, you customized, added information to, and even built graphics. You should now be able to:

- Customize several charts, graphs, and plots using arguments specific to each
- Use graphics functions to add information to any visual
- Create custom graphics by building them from the ground up

Armed with a sound strategy, talented and loyal generals, and the emperor's approval, you are ready for battle. You have come a long way since the legendary Zhuge Liang's death thrust the fate of the Shu kingdom into your hands. Your mastery of R has grown tremendously and will continue to aid you in conducting data analyses. While the future of the Shu kingdom may be uncertain, your talents are unquestioned and your knowledge will continue to blossom.

In Chapter 10, we will focus on the future. We will look at the ways in which you excel beyond the teachings of master Zhuge Liang, and the boundaries of this book, to continually refine and expand your understanding of R.

10

Becoming a Master Strategist

Throughout this book, you have continually refined and expanded your understanding of R. We began by examining the components of the R console and how to use them effectively. We then used variables, functions, and models to organize, analyze, and assess our data. We concluded by taking an in-depth look at R's graphical capabilities.

In this, our final chapter, we will explore several ways in which we can continue to learn about R. Just because we have completed this book and acquired the skills of the legendary Zhuge Liang, does not mean that our journey is complete. There are virtually an unlimited number of topics to discover in R. This list is growing day by day, as users continually expand R's functionality. To find new R knowledge, we will focus on the learning resources that are built into R and those that can be found online.

By the end of this chapter, you will be able to do the following:

- Use R's built-in help system
- Install packages that expand R's functionality
- Take advantage of electronic learning resources, such as websites, blogs, and online communities

R's built-in resources

R has two primary built-in resources that allow us to expand our use of the software. The first is the `help()` function, which can be used to learn about various R topics. The second is R's ability to be extended through the use of user-created packages. We will explore both of these resources in detail.

Time for action – using R's help function

R has a convenient `help(...)` command that yields overview information about nearly any feature. Let us review this function:

1. Open R.

2. Execute the `help(...)` function without any arguments:

```
> #learn more about the help command by using the help(...)
function without any arguments
> help()
```

3. The R Help window will open to display documentation on the `help(...)` function.

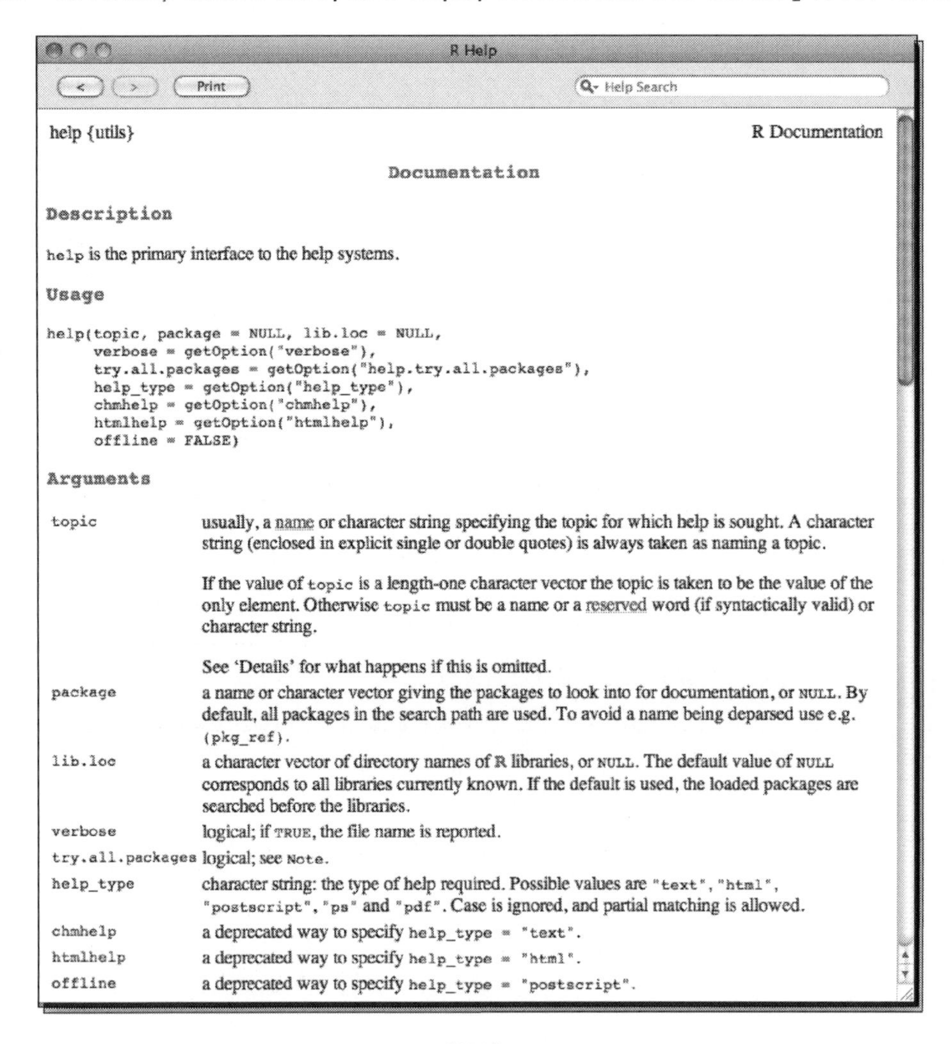

help {utils} R Documentation

Documentation

Description

help is the primary interface to the help systems.

Usage

```
help(topic, package = NULL, lib.loc = NULL,
     verbose = getOption("verbose"),
     try.all.packages = getOption("help.try.all.packages"),
     help_type = getOption("help_type"),
     chmhelp = getOption("chmhelp"),
     htmlhelp = getOption("htmlhelp"),
     offline = FALSE)
```

Arguments

topic	usually, a name or character string specifying the topic for which help is sought. A character string (enclosed in explicit single or double quotes) is always taken as naming a topic. If the value of topic is a length-one character vector the topic is taken to be the value of the only element. Otherwise topic must be a name or a reserved word (if syntactically valid) or character string. See 'Details' for what happens if this is omitted.
package	a name or character vector giving the packages to look into for documentation, or NULL. By default, all packages in the search path are used. To avoid a name being deparsed use e.g. (pkg_ref).
lib.loc	a character vector of directory names of R libraries, or NULL. The default value of NULL corresponds to all libraries currently known. If the default is used, the loaded packages are searched before the libraries.
verbose	logical; if TRUE, the file name is reported.
try.all.packages	logical; see Note.
help_type	character string: the type of help required. Possible values are "text", "html", "postscript", "ps" and "pdf". Case is ignored, and partial matching is allowed.
chmhelp	a deprecated way to specify help_type = "text".
htmlhelp	a deprecated way to specify help_type = "html".
offline	a deprecated way to specify help_type = "postscript".

4. Execute the `help(...)` function using the `topic` argument:

```
> #learn more about a specific subject using the help(...)
function with a single argument
> #the argument should specify the name of the subject that you
are seeking help on
> help(library)
```

5. The **R Help** window will open to display documentation on the specified topic:

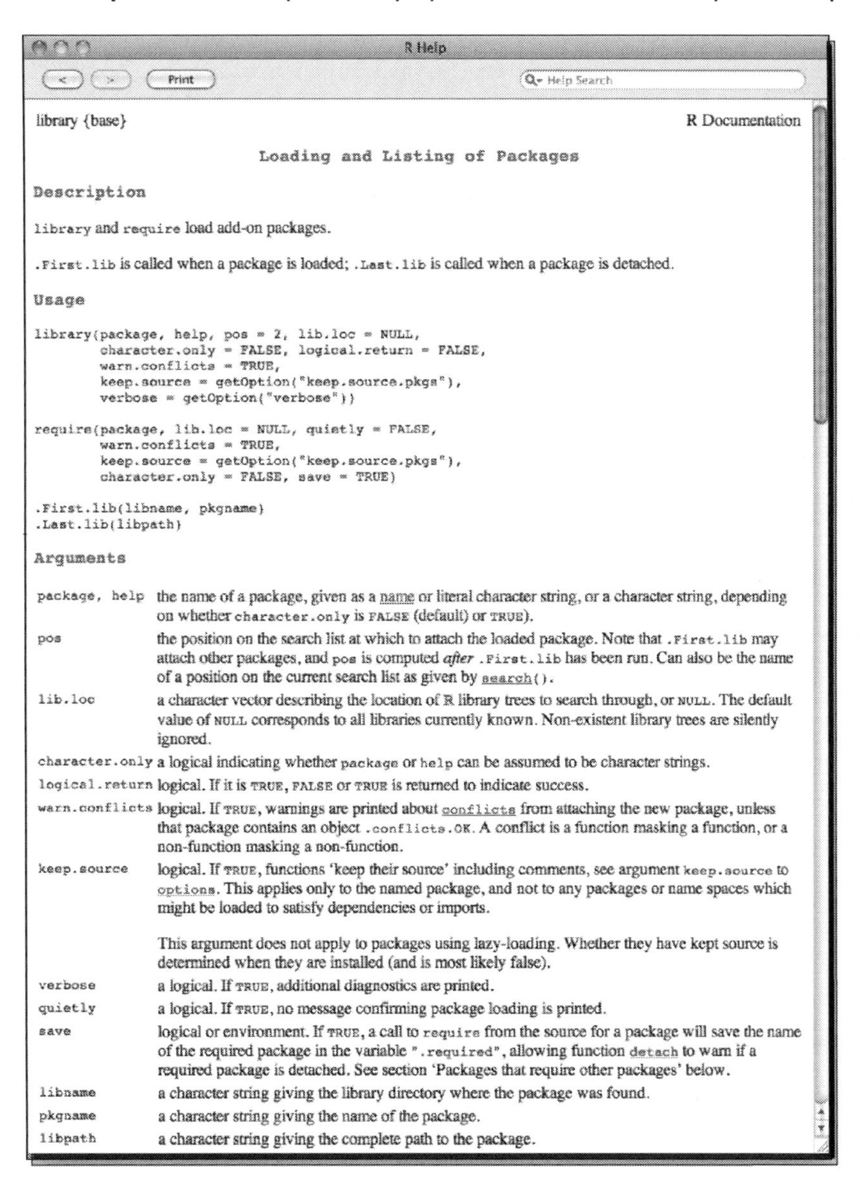

R Help

Print

Help Search

library {base} R Documentation

Loading and Listing of Packages

Description

`library` and `require` load add-on packages.

`.First.lib` is called when a package is loaded; `.Last.lib` is called when a package is detached.

Usage

```
library(package, help, pos = 2, lib.loc = NULL,
        character.only = FALSE, logical.return = FALSE,
        warn.conflicts = TRUE,
        keep.source = getOption("keep.source.pkgs"),
        verbose = getOption("verbose"))

require(package, lib.loc = NULL, quietly = FALSE,
        warn.conflicts = TRUE,
        keep.source = getOption("keep.source.pkgs"),
        character.only = FALSE, save = TRUE)

.First.lib(libname, pkgname)
.Last.lib(libpath)
```

Arguments

`package, help`	the name of a package, given as a name or literal character string, or a character string, depending on whether `character.only` is `FALSE` (default) or `TRUE`).
`pos`	the position on the search list at which to attach the loaded package. Note that `.First.lib` may attach other packages, and `pos` is computed *after* `.First.lib` has been run. Can also be the name of a position on the current search list as given by `search()`.
`lib.loc`	a character vector describing the location of R library trees to search through, or `NULL`. The default value of `NULL` corresponds to all libraries currently known. Non-existent library trees are silently ignored.
`character.only`	a logical indicating whether `package` or `help` can be assumed to be character strings.
`logical.return`	logical. If it is `TRUE`, `FALSE` or `TRUE` is returned to indicate success.
`warn.conflicts`	logical. If `TRUE`, warnings are printed about `conflicts` from attaching the new package, unless that package contains an object `.conflicts.OK`. A conflict is a function masking a function, or a non-function masking a non-function.
`keep.source`	logical. If `TRUE`, functions 'keep their source' including comments, see argument `keep.source` to `options`. This applies only to the named package, and not to any packages or name spaces which might be loaded to satisfy dependencies or imports. This argument does not apply to packages using lazy-loading. Whether they have kept source is determined when they are installed (and is most likely false).
`verbose`	a logical. If `TRUE`, additional diagnostics are printed.
`quietly`	a logical. If `TRUE`, no message confirming package loading is printed.
`save`	logical or environment. If `TRUE`, a call to `require` from the source for a package will save the name of the required package in the variable `".required"`, allowing function `detach` to warn if a required package is detached. See section 'Packages that require other packages' below.
`libname`	a character string giving the library directory where the package was found.
`pkgname`	a character string giving the name of the package.
`libpath`	a character string giving the complete path to the package.

What just happened?

We demonstrated how to use the `help(...)` function to learn about R components using its built-in documentation system.

help(...)

Our first use of `help(...)` contained no arguments and therefore conveniently returned documentation on the `help(...)` function itself. Most often, you will want to use the `help(...)` function in coordination with a single argument that specifies the subject that you are seeking help on. For instance, we used the `help(...)` function to learn more about the `library(...)` function:

```
> help(library)
```

Notice that the argument in the `help(...)` function is simply the name of a topic. As demonstrated, when the topic is a function, no parenthesis, ellipsis, or arguments should be included with the function name.

When executed, the `help(...)` function opens the **R Help** window to display the documentation related to the identified subject. The **R Help** window will display a brief description of the topic along with sections explaining its usage, arguments, details, author(s), references, examples, and related subjects. All of R's help documentation follows this format, although each individual help page may not contain every section.

The `help(...)` function is a fast and easy way for experienced users to retrieve information. It is best for users who are already familiar with specific R topics and need to be reminded of certain arguments or features. It is not always an optimal tool for learning how to do brand new things, but the built-in help system is the official resource for R documentation.

Pop quiz

1. What happens when the `help()` function is executed without any arguments?
   a. It fails to execute and returns an error.
   b. The **R Help** window displays documentation on the specified subject.
   c. The **R Help** window displays documentation on the `help(...)` function.
   d. The **R Help** window displays a menu of selectable help topics.
2. What happens when the `help(...)` function is executed with a subject argument specified?
   a. It fails to execute and returns an error.
   b. The **R Help** window displays documentation on the specified subject.

c. The **R Help** window displays documentation on the `help(...)` function.

d. The **R Help** window displays a menu of selectable help topics.

Have a go hero

Review the help documentation on the `chooseCRANmirror()`, `install.packages()`, and `library(...)` functions to prepare for the next section on packages.

Time for action – expanding R with packages

R's functionality can be easily and significantly expanded through the use of **packages**. A package is a collection of functions that has been contributed by members of the R user community. Let us look at the steps involved in acquiring, installing, loading, and using a new package in R:

1. Open the CRAN mirror window using the `chooseCRANmirror()` command:

```
> #acquiring, preparing, installing, and using a new R package
> #step 1: choose a CRAN mirror
> #open the CRAN mirror window using chooseCRANmirror()
> #then choose the mirror located nearest to you
> chooseCRANmirror()
```

A new window will open to display the available CRAN mirrors. Choose the mirror that is located nearest to you, then click on the **Ok** button:

Note that the appearance of the CRAN mirror window may vary depending on the operating system and version that you use.

2. Open the packages window using the `install.packages()` command:

```
> #step 2: install the package
> #open the packages window using install.packages
> #then choose a package to install it on your computer
> install.packages()
```

A new window will open to display the available packages. Choose the **magic** package, then click on the **Ok** button.

Note that the appearance of the packages window may vary depending on the operating system and version that you use. Also note that R will automatically install any packages that the selected package depends upon to operate.

3. Use the `library(...)` function to load the a package for use in R.

```
> #step 3: load the package
> #use the library(...) function to load a package once it has been installed
> #load the magic package
> library(magic)
```

Note that R will automatically load any packages that the specified package depends upon to operate and notify you in the console. If no additional packages are necessary, R will drop down to the next line without providing any output.

4. R will drop down to the next line. The package is now ready to be used.

5. Use the `magic(n)` function from the magic package to generate a sequence of magic squares:

```
> #step 4: use the package
> #once loaded, a package's functions can be used within the R console
> #use the magic(n) function from the magic package to generate an 8x8 magic square
> #n is a single nonnegative number that indicates how many rows and columns the magic square will have
> #this function generates the same type of magic square that we encountered when solving Zhuge Liang's puzzle in chapter 3!
> magic(8)
```

```
> magic(8)
     [,1] [,2] [,3] [,4] [,5] [,6] [,7] [,8]
[1,]    1   56   48   25   33   24   16   57
[2,]   63   10   18   39   31   42   50    7
[3,]   62   11   19   38   30   43   51    6
[4,]    4   53   45   28   36   21   13   60
[5,]    5   52   44   29   37   20   12   61
[6,]   59   14   22   35   27   46   54    3
[7,]   58   15   23   34   26   47   55    2
[8,]    8   49   41   32   40   17    9   64
```

What just happened?

We expanded the capabilities of R by downloading, installing, and loading a package in the R console. Let us review the steps involved in this process.

Choose a CRAN mirror

The initial step in acquiring a new R package is to choose a **CRAN mirror**. The acronym **CRAN** stands for **Comprehensive R Archive Network** and refers to several worldwide servers that store and maintain R's code and documentation. A CRAN mirror is a single server in this network. When choosing a CRAN mirror, it is best to select the location nearest to you. Since the data that you request will travel a shorter distance, you will be able to download more content in less time. Using the `chooseCRANmirror()` function will open the **CRAN mirror** window, which displays a list of all available CRAN mirrors.

> Note that `chooseCRANmirror()` only needs to be executed once each time that you launch R. Once a CRAN mirror is selected, it will remain active until you quit R.

Install a package

Next, you will need to install the desired package. Executing the `install.packages()` command in the R console will open the **Packages** window, which displays a list of all available packages.

In our example, we selected the `magic` package. R then automatically installed the `abind` package, which is required for `magic` to function. Whenever necessary, R will automatically install required packages, known as **dependencies**, in this fashion.

> A given package only needs to be installed once. It is then available to be loaded any time that you use R.
>
> Also note that if you already know the name of the package, you can install it using a single `install.packages(name)` command, such as `install.packages("magic")`.
>
> A list of every available R package, along with a description of each, can be found on the official R website at:
>
> `http://cran.r-project.org/web/packages`

Load the package

Then, to prepare the package for use in the R console, it must be loaded via the `library(...)` function. This function receives an argument that indicates the name of the function. For instance, since we wanted to load the `magic` package in our activity, the `library(...)` function took on the following form:

```
> library(magic)
```

As with `install.packages()`, the `library(...)` command will automatically load any necessary dependencies. In our case, the `abind` package was loaded after executing our `library(...)` function. When no dependencies are present, R will drop down to the next line in the console without providing any output.

A given package only needs to be loaded once each time that you launch R. Once loaded, a package will remain active until you quit R.

Also note that you can check for and install updates to your R packages using the `update.packages()` command.

Use the package

Once you have loaded a new package in R, the final step is to take advantage of its offerings. Quite simply, once a package has been loaded, you can use any of its functions, just as we have been using R's built-in functions throughout this book.

In our activity, we loaded the magic package, which gave us access to several functions related to magic squares. We used the `magic(n)` function to generate an 8x8 magic square.

This is the same variety of magic square that we encountered when solving Zhuge Liang's puzzle in *Chapter 3*. In fact, the puzzle that you solved in that chapter was generated using R and the `magic` package!

```
> magic(8)
     [,1] [,2] [,3] [,4] [,5] [,6] [,7] [,8]
[1,]    1   56   48   25   33   24   16   57
[2,]   63   10   18   39   31   42   50    7
[3,]   62   11   19   38   30   43   51    6
[4,]    4   53   45   28   36   21   13   60
[5,]    5   52   44   29   37   20   12   61
[6,]   59   14   22   35   27   46   54    3
[7,]   58   15   23   34   26   47   55    2
[8,]    8   49   41   32   40   17    9   64
```

All R packages can be installed by following this same procedure. The immense value of R packages is that they expand the capabilities of R. Thousands of packages are currently available and new ones are continuously being created by the R user community. This means that R is perpetually growing in scope and functionality.

Pop quiz

1. How often must the `chooseCRANmirror()` function be executed in R?
   a. Once.
   b. Once each time R is launched.
   c. Once each time a given package is installed.
   d. Once each time a given package is loaded.
2. How often must the `install.packages()` function be executed in R?
   a. Once.
   b. Once each time R is launched.
   c. Once each time a given package is installed.
   d. Once each time a given package is loaded.
3. How often must the `library(...)` function be executed in R?
   a. Once.
   b. Once each time R is launched.
   c. Once each time a given package is installed.
   d. Once each time a given package is loaded.

Have a go hero

Use the R website's online package listing (`http://cran.r-project.org/web/packages`), or one of the other resources presented in this chapter, to learn about the packages that are available in R. Choose one that will be useful to your work. Then install it in R and experiment with its functions.

R's online resources

A wealth of online resources are available for R. These include search engines, websites, blogs, and online communities. Some of the most useful and informative online resources for learning about R will be discussed here.

Websites

A few valuable R websites are highlighted here.

The R Project for Statistical Computing

The official R website is the definitive source for R updates and documentation. It offers the most recent versions of R for each operating system. Also included are valuable documents, such as the R FAQ, CRAN mirror listing, and contributed packages glossary. The **R Project for Statistical Computing** can be found at:

```
http://www.r-project.org
```

Quick-R

Quick-R is an excellent resource for efficiently retrieving information on R topics, along with examples of how related techniques can be executed. It covers a wealth of subjects in R, including a wide range of statistical methods and graphics types. Its organized and aesthetic format makes it easy to locate and decipher the desired information. **Quick-R** can be found at:

```
http://www.statmethods.net
```

R Programming wikibook

The **R Programming** wikibook shares information on general R topics, as well as references to statistical methods. All of its content is presented in wiki format with minimal description, making it a resource for quickly locating and indulging in code samples. The **R Programming** wikibook can be found at:

```
http://en.wikibooks.org/wiki/R_Programming
```

R Graph Gallery

The **R Graph Gallery** presents a collection of some of the most advanced and unique graphics that have been generated using R. Images of each visualization are accompanied by the source code used to create them and references to any required packages. The **R Graph Gallery** can be found at:

```
http://addictedtor.free.fr/graphiques
```

Crantastic!

Crantastic! is a website dedicated to R packages. It features an up-to-date searchable listing of all R packages, along with their descriptions, web addresses, version information, and author details. **Crantastic!** can be found at:

```
http://crantastic.org
```

Blogs

Blogs are another informative online resource for learning about R.

R bloggers

R bloggers is an extensive collection of over 100 blogs (and counting) that are dedicated to sharing knowledge related to R. It is a prime resource for gaining insights on complex and cutting edge data analysis techniques. The combined authorship of **R bloggers** represents perhaps the most active and timely sharers of R content on the internet. **R bloggers** can be found at:

```
http://www.r-bloggers.com
```

R Tutorial Series

The **R Tutorial Series** provides user-friendly guides for people who are learning about R. It primarily focuses on providing brief statistics tutorials with detailed execution examples. This blog was created by the author of this book and follows a similar style. The **R Tutorial Series** is available at:

```
http://rtutorialseries.blogspot.com
```

Online communities

Online communities are places to connect with other R users and seek and share information.

R-help mailing list

The **R-help** mailing list is both an ongoing resource for answers to R questions and an archive of past conversations dating back to 1997. The odds are that any difficulty that you encounter in R has already been queried to this list at one time or another. If not, you can always join the list and contribute to the community by submitting your own questions. Therefore, the **R-help** mailing list is one of the first places you should look when you are having trouble with a particular facet of R. You can join the *R-help* mailing list at:

```
https://stat.ethz.ch/mailman/listinfo/r-help
```

A searchable archive of the **R-help** mailing list is available at:

`http://tolstoy.newcastle.edu.au/R`

Other mailing lists

A number of additional mailing lists are available for users seeking information about R. The most prominent ones, which cover the major announcements, contributed packages, and development of R can be found using the **Mailing Lists** link on the official website at:

`http://www.r-project.org`

Search engines

Endless amounts of information about R can also be found by searching the internet.

R Seek

R Seek is a Google-based search engine that helps users find what they are looking for by automatically optimizing their queries to yield relevant results. Users can search for a topic as they normally would, but are more likely to be presented with output that is related to R than if they had entered the same terms into a standard search engine. The **R Seek** search engine is available at:

`http://www.rseek.org`

Google

Google is another valuable resource for searching R, although it does take some practice to acquire meaningful results through this engine. It can be difficult to find relevant webpages via *Google*, because R's one letter name tends to be ignored or inappropriately parsed by search engine algorithms. Thus, it is often useful to use quotations along with additional terms when searching *Google*. For example, to find information on conducting *multiple regression* in R, it is better to use include additional quoted keywords, such as **R Project** or **R statistics** than just the letter *R* alone. The **Google** search engine is available at:

`http://www.google.com`

Summary

In this chapter, you explored several resources for broadening your understanding of R. These consisted of built-in features, such as the `help(...)` command and packages, and internet resources, including websites, blogs, and online communities. You should now be able to do the following:

- Use R's built-in help system
- Install packages that expand R's functionality
- Take advantage of electronic learning resources, such as websites, blogs, and online communities

Congratulations on completing this book. You have discovered much of what R has to offer and should feel comfortable incorporating it into your everyday work. Do not cease to refine and expand your knowledge of R. It is hoped that R will be a useful tool in your work for many years to come.

On a final note, you have earned your place amongst a global group of analysts, businesspeople, academics, scientists, and others with a passion for open source data analysis software. Welcome to the R user community.

Pop Quiz Answer Key

Chapter 2

Setting your R working directory	
1	d
2	b
3	d

Chapter 3

Solving the first 4x4 Magic Square	
1	b
2	d
3	c

Chapter 4

Accessing data within variables	
1	d
2	b
3	d

Manipulating variable data	
1	10 20 30 40 50 60
2	1 12 3 4 5 6
3	c

Managing the R workspace	
1	c
2	d
3	a

Chapter 5

Creating a subset from a large dataset	
1	d
2	a

Deriving summary statistics	
1	c
2	b

Quantifying categorical variables	
1	d
2	a

Correlating variables	
1	b
2	d

Modeling with simple linear regression	
1	b
2	a
3	c

Modeling with multiple linear regression	
1	a
2	c

Modeling interactions	
1	b
2	d
3	a

Comparing and choosing models	
1	d

Chapter 6

Creating custom functions	
1	c
2	b
3	c

Incorporating resource constraints into predictions	
1	d

Assessing the viability of potential strategies	
1	d

Chapter 7

Data setup	
1	c

Data exploration	
1	c

Model development	
1	d
2	d

Model deployment	
1	c

The common steps to all R analyses	
1	d

Chapter 8

Creating a bar chart	
1	a
2	d

Customizing graphics	
1	d
2	d

Creating a scatterplot	
1	a
2	c

Creating a line chart	
1	d
2	a

Creating a box plot	
1	a
2	c

Creating a histogram	
1	d

Creating a pie chart	
1	b

Exporting graphics	
1	a

Chapter 9

Customizing a bar chart	
1	b
2	d
3	b

Customizing a scatterplot	
1	a
2	c

Customizing a line chart	
1	d
2	c

Customizing a box plot	
1	b
2	c

Customizing a histogram	
1	b
2	c

Customizing a pie chart	
1	a
2	c

Building a graphic	
1	a
2	d

Building a graphic with multiple visuals	
1	c
2	c
3	c

Chapter 10

Using R's help function	
1	c
2	b

Expanding R with packages	
1	b
2	a
3	b

Index

Symbols

.csv files 44
4x4 Magic Square puzzle
  solving 35, 36

A

abind package 260, 261
abline(...) function
  about 209
  arguments 209
  example 210
  syntax 209
  using 204, 205
abline(...) function, arguments
  lty 209
  reg 209
AIC
  about 137
  procedure 137
  using 138
AIC(object, ...) function 138
Akaike Information Criterion. *See* **AIC**
ambush model 106
analysis of variance (ANOVA)
  about 99
  interpreting 99
angle argument
  about 197, 198
  shade, changing for bar chart 192, 197
anova(object, ...) function 100, 138
as.numeric(data) function
  about 75
  example 75
attach(variable) function 48, 49
axis(...) function
  custom axes, creating for box plot 222-224
  using 223, 224

B

bar chart
  about 152
  creating, in R 152, 153
  customizing 186-193
  legend, adding 193
  text labels, adding 194
bar chart, customizing
  col argument 159, 161
  main argument 159, 160
  xlab argument 159, 160
  xlim argument 159, 160
  ylab argument 159, 160
  ylim argument 159, 160
bar graph. *See* **bar chart**
barplot(...) function
  about 153-186
  arguments 194, 195
  bar chart, creating 152, 153
barplot(...) function, arguments
  horiz 195, 196
  names 194
  space 194, 195
  width 194, 195
battle data
  importing, into R 126, 127
battle plan
  selecting 123
battle strategy
  logistical considerations 117, 118

resource map 118
viability, assessing 121, 122
beside argument
about 196
stacked bar chart, creating 190, 191, 197
blogs
about 264
R bloggers 264
R Tutorial Series 264
box plot
creating 172-174
customizing 220
boxplot(...) function
about 174
box plot, creating 172-174
breaks argument
histogram's columns, separating 225
using 228
built-in resources, R
about 253
help(...) command 254-256
packages 257-259

C

c(...) function 154, 156
calculations
about 38
performing, on cell 54
performing, on column 54
performing, on dataset 53, 54
performing, on row 54
categorical variables
quantifying 73, 74
causation 79
cell
calculations, performing on 54
centering
about 94
need for 94
cex argument
about 198, 206
scatterplot's point markers, customizing 199
chooseCRANmirror() command
CRAN mirror window, opening 257
using 260
coef(object) command
about 143
using 143
col argument
about 159, 161
rainbow colors, generating 161
specific colors, generating 161, 162
column
calculations, performing on 54
comma-separated values files. *See* **.csv files**
comment 37, 38
Comprehensive R Archive Network. *See* **CRAN**
content, R workspace
listing 58
loading 59
saving 59
cor() function 78
cor(data) function
about 80
limitations 80
using 80
cor(x,y) function 79
correlations
about 77, 78
interpreting 78, 79
correlations, interpreting
about 78
examples 79
count 228
CRAN 21, 260
CRAN mirror
about 260
selecting 260
Crantastic!
about 264
URL 264
custom functions
creating 111, 112, 142
example 113

D

data
accessing, within variables 47-50
analyzing 126
examining 65
exploring 129-132, 147

external data, importing 43, 44
importing 146
initial reference, making 63, 64
setting up 126
dataset
calculation, performing on 53, 54
subset, creating 66
date() command 30, 67
density 228
density argument
about 197, 198
shade, changing for bar chart 192, 197
dependencies 260
descriptive statistics. *See* **summary statistics**
detach(variable) function 50
dichotomous 211
dir argument 31

E

equivalency operator 67
extended lines
about 114
formatting value 114
external data
importing 43, 44

F

fire model 107
freq argument
densities, plotting 226, 227
using 228
function() command
about 111, 113
syntax 113
using 111, 112, 140
function arguments
variable data, using 54

G

getwd() command 31, 126
glm(formula, data) function
about 138
using 132
Google
about 265
URL 265
graphic device
creating 182
graphics
about 249
axes, adding 235, 240
creating 183, 234, 235
creating, with multiple visuals 242-247
customizing 156-158
data, adding 236, 237, 240
exporting 181, 182
horizontal lines 250
legend, adding 238, 241
nested functions 250
vertical lines 250
graphic window
about 154
graphics, customizing 156-158
multiple graphics, displaying 242
Quartz 154
working 154, 155

H

head to head model 104
height argument 153
help(...) command
about 256
executing 256
executing, topic argument used 255
features 256
using 254-256
hierarchical linear regression (HLR)
models, comparing 96-98
hist(...) function
about 176
histogram, creating 175, 176
histogram
about 175
creating 175, 176
customizing 225
horiz argument
about 195
bar chart orientation, changing 188, 189
working 195, 196

I

initial reference
making, for data 63, 64
install.packages() command
package window, opening 258, 260
installation, R
about 20
CRAN link, clicking 21
CRAN link, displaying 22
installation, R-2.11.1.pkg
Mac OS X 10.5, used 24-28
interaction effects 92
interaction variable
about 92
creating 92-94
incorporating, in regression model 93, 94
interpreting 94
interquartile range 223

L

labels argument
percentages, adding to pie chart 230
legend
adding, to bar chart 193, 198
adding, to graphics 238, 241
adding, to pie chart 231
adding, to scatterplot 203, 204
legend(...) function
about 162, 198
arguments 162
example 209, 233
format 162
using 219
legend(...) function, arguments
fill 162
legend 162
x position 162
y position 162
length(object) command 161
library(...) function
package, loading 259, 260
linear regression model
interpreting 86, 87
line chart
about 168
additional data, incorporating 213
creating 168, 169
customizing 212
displaying, in graphic window 213
legend, adding 215
lines(...) function, using 214
lwd argument, using 216
line graph. *See* **line chart**
lines 37
lines(...) function
about 217
arguments 217
example 217
syntax 217
using 214
lines(...) function, arguments
col 217
type 217
x 217
y 217
lm(formula, data) function 84, 138
load(file) function 58, 59, 63
logistics
considering 117
logistics, considering
about 117
equipment 118
gold 117
provisions 117
soldiers 118
ls() function 57, 58
lwd argument
line width, setting 212
using, in line chart 216

M

Mac OS X 10.5
R-2.11.1.pkg, installing 24-28
magic(n) function
magic squares sequence, generating 259, 261
main argument 159, 160
matrix(...) function
about 197
format 197
mean(data) function 54
meanDurationAmbush variable 110
meanDurationFire variable 110

meanDurationHeadToHead variable 110
meanDurationSurround variable 110
model
deplolying 139-141
developing 132-135
model summaries
interpreting 98
moderation effect 94
multi-argument function 67
multicollinearity 94
multiple linear regression
about 88
creating 88
modelling with 89

N

names argument
text labels, adding to bar chart 194
non-equivalency operator 67
number-colon-number notation
about 170
benefits 171

O

online communities
R-help mailing list 264
online resources, R
blogs 264
online communities 264
websites 263
output 38

P

packages
about 257
CRAN mirror, selecting 260
installing 260
loading 260
using 261
par(mfcol) command
about 249
using 249
paste(...) function
about 232
arguments 232
syntax 232
paste(...) function, arguments
appendText 232
originalValues 232
sep 232
pch argument
about 206
scatterplot's point markers, customizing 199
pie(...) function
about 179
pie chart, creating 177-179
pie chart
about 177
creating 177-179
customizing 230
legend, adding 231
plot(...) function
about 170
line chart, creating 168, 169
relationship, exploring among dataset 167
scatterplot, creating 164-166
type argument 170
using 234, 239
points(...) function
about 207
arguments 207
executing 207, 209
relationships, adding 202, 203
syntax 207
points(...) function, arguments
col 207
type 207
x 207
y 207
predictions
resource constraints, incorporating into 119
probabilitySuccessAmbush variable 109
probabilitySuccessFire variable 109
probabilitySuccessHeadToHead variable 109
probabilitySuccessSurround variable 109
probability values
calculating 108

Q

q() command 57
Quartz 154
Quick-R
about 263
URL 263

R

R
4x4 Magic Square puzzle, solving 35, 36
about 8
bar chart, creating 152, 153
bar chart, customizing 186-193
battle data, importing 126, 127
benefits 8, 9
commands, issuing 29
data, accessing within variables 47-50
expanding, with packages 257-259
external data, importing 43, 44
function() command 111, 112
graphics, exporting 181, 182
help(...) command, using 254-256
homepage, URL 20
initial reference, making for data 63, 64
need for 9
quitting 59
regression models 82, 103
URL 8
variables, calling 45, 46
variables, creating 45, 46
working directory, setting 30, 63
workspace, managing 57, 58
Zhuge Liang's magic square, deciphering 34
R, analyzing
about 145
console files, saving 148, 149
data, exploring 147
data, importing 146
working directory, setting up 145
workspace, saving 148, 149
R, benefits
about 8
community-supported 9
cross-platform 8
extendable 9
free 8
graphical 9
open source 8
programmable 9
R, fundamental elements
calculations 38
comment 37, 38
lines 37
output 38
R, installing
about 20
CRAN link, clicking 21
CRAN link, displaying 22
R-2.11.1.pkg, installing
Mac OS X 10.5, used 24-28
R-help mailing list
about 264
URL 264
rainbow(...) function 161, 162
range(data) function 72
range argument
about 223
whisker length, altering for box 220
Rating value 108
ratioWeiShuSoldiersAmbush variable 109
ratioWeiShuSoldiersFire variable 109
ratioWeiShuSoldiersHeadToHead variable 109
ratioWeiShuSoldiersSurround variable 109
R bloggers
about 264
URL 264
R command
issuing 29
R console
saving 60
versus R workspace 59, 60
visualizing 39
read.csv(file) command
about 44
resource file, reading into R 44
read.table(...) function
about 128
arguments 128
read.table(...) function, arguments
file 128
header 128
sep 128

regression equation
format 84
regression line 209
regression models
about 82, 103
ambush 106
comparing 96-98
fire 107
head to head 104
outcomes, calculating from 110, 111
probability values, calculating 108
selecting 96-98
simple linear regression 82, 83
surround 105
resource-focused custom functions
creating 115, 116
resource constraints
incorporating, into predictions 119
resource map 118
R Graph Gallery
about 263
URL 263
R Help window 255
round(x, digits) function 231
row
calculations, performing on 54
R Programming wikibook
about 263
URL 263
R Project for Statistical Computing
about 263
URL 263
R Seek
about 265
URL 265
R Tutorial Series
about 264
URL 264
R workspace
contents, listing 58
contents, loading 59
contents, saving 59
managing 57, 58
saving 148, 149
versus R console 59, 60

S

save.image(file) function 57, 59, 145, 148
scatterplot
about 164, 206
additional data, incorporating 200, 201
creating 164-166
customizing 199
displaying, in graphic window 201
legend, adding 203, 204
multiple scatterplot 167
points, adding 202, 203
single scatterplot 167
sd(data) function
about 71
using 71
search engines
about 265
Google 265
R Seek 265
setwd(dir) function 31, 43, 126
simple linear regression
about 82
modelling with 82, 83
space argument 194
standard deviation 71
subset
creating, from dataset 66
subset(data, ...) function 66, 67
SuccessfullyExecuted variable 105
summary(object) function 72
about 147
using 85, 131, 132, 144
summary output
interpreting 90
p-value 86
R-squared 86
summary statistics
deriving 69-71
examining 129-132
need for 72
sunflowerplot(...) function
using 211
surround model 105

T

topic argument
help(...), executing 255
type argument 170

U

update.packages() command 261

V

variable$column notation 49
variable-argument function
about 67
anova(object, ...) 100
variable[row, column] notation 50
variable data
manipulating 51, 52
using, in function arguments 54
variables
calling 45, 46
categorical variables, quantifying 73, 74
correlating 77, 78
creating 45, 46
data, accessing within 47-50
interaction variables 92
overwriting 75, 76
variable calculation, saving 55
vector variable 154
viability
assessing 121, 122

W

websites
Crantastic! 264
Quick-R 263
R Graph Gallery 263
R Programming wikibook 263
R Project for Statistical Computing 263
width argument 194
working directory
about 30
setting 30, 145

X

xlab argument 159, 160
xlim argument 159, 160, 194

Y

ylab argument 159, 160
ylim argument 159, 160

Z

Zhuge Liang 7
Zhuge Liang's magic square puzzle
about 34
deciphering 34

Thank you for buying

Statistical Analysis with R Beginner's Guide

About Packt Publishing

Packt, pronounced 'packed', published its first book "*Mastering phpMyAdmin for Effective MySQL Management*" in April 2004 and subsequently continued to specialize in publishing highly focused books on specific technologies and solutions.

Our books and publications share the experiences of your fellow IT professionals in adapting and customizing today's systems, applications, and frameworks. Our solution based books give you the knowledge and power to customize the software and technologies you're using to get the job done. Packt books are more specific and less general than the IT books you have seen in the past. Our unique business model allows us to bring you more focused information, giving you more of what you need to know, and less of what you don't.

Packt is a modern, yet unique publishing company, which focuses on producing quality, cutting-edge books for communities of developers, administrators, and newbies alike. For more information, please visit our website: `www.packtpub.com`.

About Packt Open Source

In 2010, Packt launched two new brands, Packt Open Source and Packt Enterprise, in order to continue its focus on specialization. This book is part of the Packt Open Source brand, home to books published on software built around Open Source licences, and offering information to anybody from advanced developers to budding web designers. The Open Source brand also runs Packt's Open Source Royalty Scheme, by which Packt gives a royalty to each Open Source project about whose software a book is sold.

Writing for Packt

We welcome all inquiries from people who are interested in authoring. Book proposals should be sent to author@packtpub.com. If your book idea is still at an early stage and you would like to discuss it first before writing a formal book proposal, contact us; one of our commissioning editors will get in touch with you.

We're not just looking for published authors; if you have strong technical skills but no writing experience, our experienced editors can help you develop a writing career, or simply get some additional reward for your expertise.

Catalyst 5.8: the Perl MVC Framework

ISBN: 978-1847199-24-9 Paperback: 244 pages

Build scalable and extendable web applications using the agile MVC framework

1. Increase reusability and empower the delivery of more complex design patterns by extending the MVC concept
2. Build an editable web interface
3. Extend Catalyst through plugins
4. Plenty of examples with detailed walkthroughs to create sample applications
5. Updated for the latest version, Catalyst 5.8

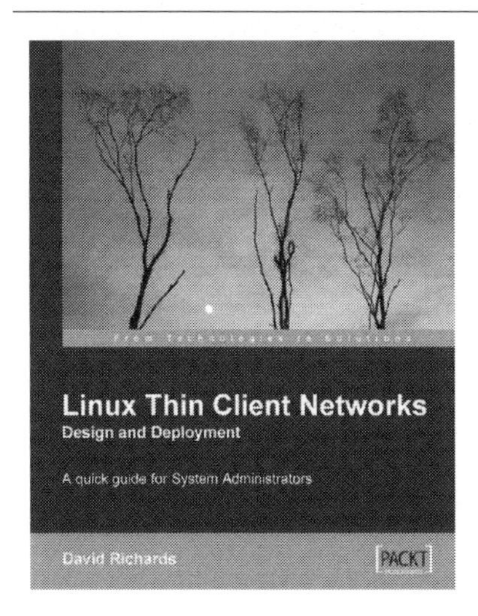

Linux Thin Client Networks Design and Deployment

ISBN: 978-1-847192-04-2 Paperback: 176 pages

A quick guide for System Administrators

1. Learn to implement the right Linux thin client network for your requirements
2. Evaluate and choose the right hardware and software for your deployment
3. Techniques to intelligently design and set up your thin client network
3. Practical advice on educating users, convincing management, and intelligent use of legacy systems

Please check **www.PacktPub.com** for information on our titles

Printed in Great Britain
by Amazon.co.uk, Ltd.,
Marston Gate.